Gerard's Arboral

TREES FROM

THE GREAT HERBALL

or

GENERALL HISTORIE OF PLANTES

containing the description, place, times, names, natures & vertues

OF ALL SORTS OF TREES

for meate, medicine, or sweet-smelling use & for use in diverse trades on land and sea, &c.

GATHERED BY

JOHN GERARD

Master of Surgery

LONDON

ENLARGED & AMENDED BY

THOMAS JOHNSON

Citizen and Apothecary of London

VELLUMINOUS PRESS

anno 2012

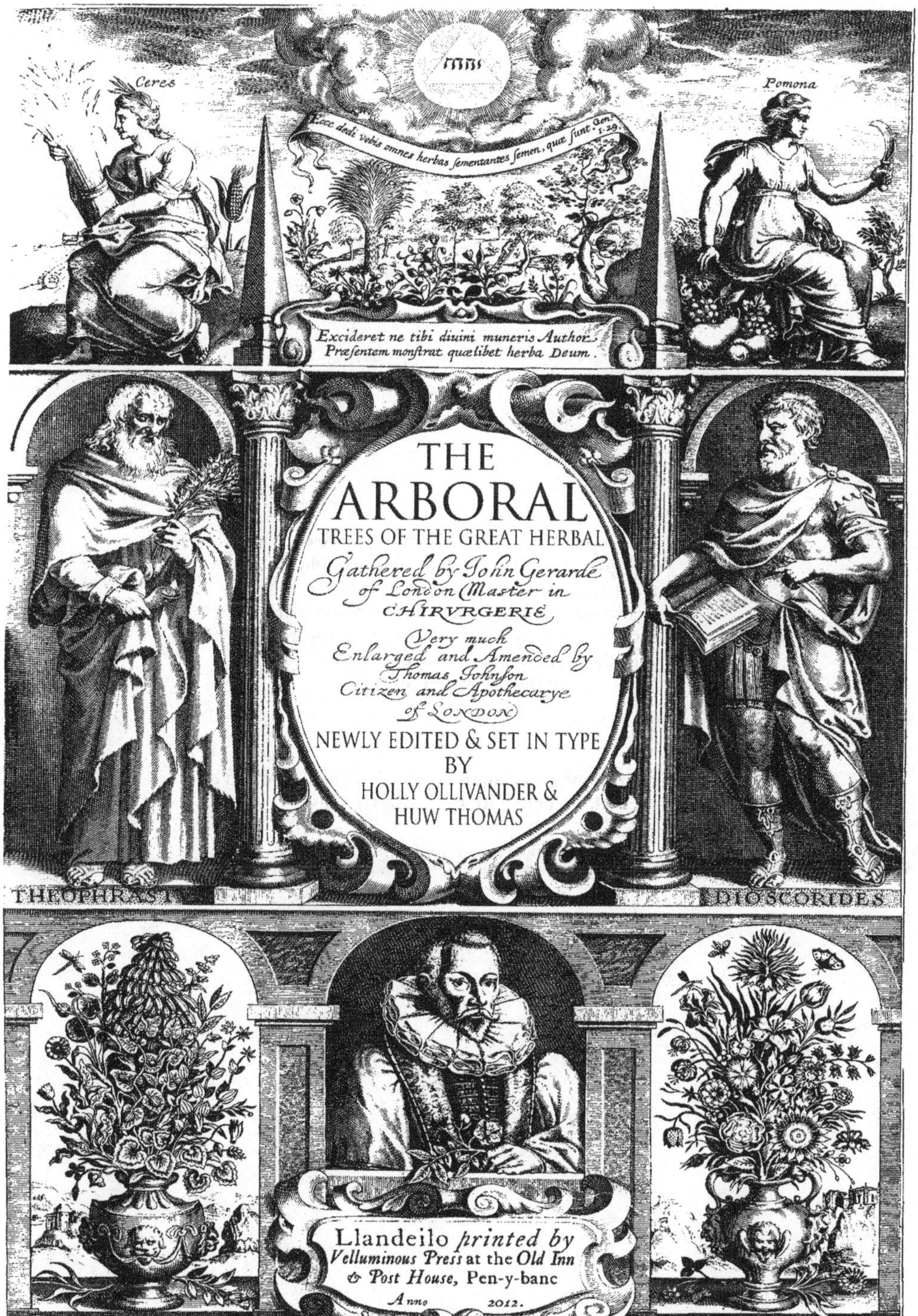

Ceres
Pomona
Ecce dedi vobis omnes herbas sementantes semen, quæ sunt. Gen. 1.29.
Excideret ne tibi diuini muneris Author
Præsentem monstrat quælibet herba Deum.
THE
ARBORAL
TREES OF THE GREAT HERBAL
Gathered by John Gerarde of London Master in CHIRURGERIE
Very much Enlarged and Amended by Thomas Johnson Citizen and Apothecarye of LONDON
NEWLY EDITED & SET IN TYPE BY
HOLLY OLLIVANDER & HUW THOMAS
THEOPHRAST
DIOSCORIDES
Llandeilo printed by
Velluminous Press at the Old Inn
& Post House, Pen-y-banc
Anno 2012.

Translated from the 1633 manuscript, designed & edited by
Holly Ollivander & Huw Thomas

1633 frontispiece by John Payne

ISBN: 978-1-905605-31-6
Velluminous Press
www.velluminous.com

Gerard's Arboral

"The honey is too dear that is lick't from thorns."

Contents

Publisher's Note

The material contained herein dates from the Elizabethan age; Gerard's advice, period recipes, and so on, are not intended to replace professional medical or nutritional advice.

Preface

As the 1500s unfolded, translations began to be undertaken of the great *Materia Medica* of the Greek physician Dioscorides of Anazarba. In 1530, Otto Brunfelsius originated the first botanical woodcuts to accompany herbal commentaries. In 1554, the Herbal of Rembertus Dodonaeus was published to great acclaim in Dutch. Henry Lyte's English translation followed in 1578, and five years later a certain Dr Priest translated the entire botanical works of Dodonaeus with even greater detail. It was this translation of Priest's that served as the acorn of what was to become the mighty oak nurtured by John Gerard.

For the educated gentleman funded by a noble patron and free to single-mindedly pursue his chosen passion, the latter decades of the 16th century were a wonderful time to be alive and the meticulous, exhaustive collection of Gerard's Great Herbal is the fruit of this freedom.

The Elizabethan world had been transformed by the discovery of what were believed to be the long-sought territories of the eastern edge of Asia. The chiefest treasure which poured forth from this new land was not gold from the fabled mines promised by early explorers but rather from nature herself: cotton, tobacco, the Virginian potato and sweet corn. Each year brought new and exciting discoveries, many of them carefully received by Gerard and raised, researched and recorded forever in the text and woodcuts of his Herbal, at last published in folio in 1597.

Three decades later, the celebrated botanist Thomas Johnson took on the challenge of the folio and greatly amended and enlarged many of Gerard's entries. The result was published in 1633 and stands today as a mighty block of a book; word-packed, weighty, and endlessly fascinating to explore.

It is necessary to mention one other important contributor, Marcus Woodward. His 1927 gathering focussed primarily upon plants one can see growing in hedgerows and fields on a casual ramble through the English countryside and we believe his choices were indicative of his time; in the respite between two World Wars England celebrated a resurgence of love for the botanical bounty that the average Briton might enjoy in his own land. Woodward only briefly touches upon a few foreign oddities and though it was his book that first whetted our appetites, one doesn't fill up on the starter and miss the main course. We were more than ready to sample the mighty feast that is the Great Herbal.

Preparing this gathering of trees from Gerard's and Johnson's great work proved a continually unfolding delight and our hope is that it brings you just as much pleasure, for throughout we have endeavoured *to make a vertue of necessitie* and think it a good book.

Holly Ollivander
Huw Thomas

Of Adams Apple tree, or the West-Indian Plantaine.

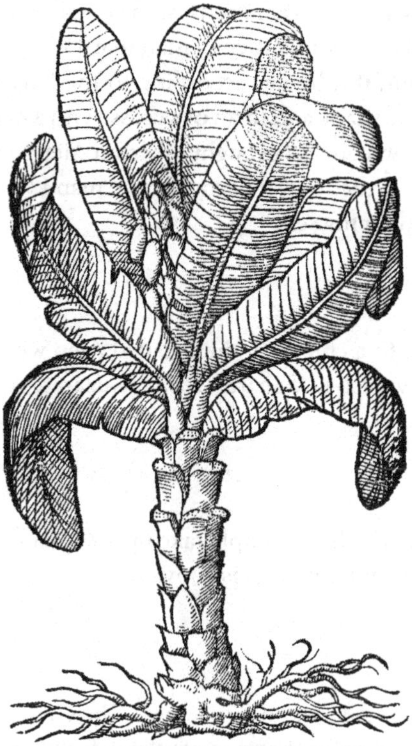

musa serapionis.
Adams Apple tree.

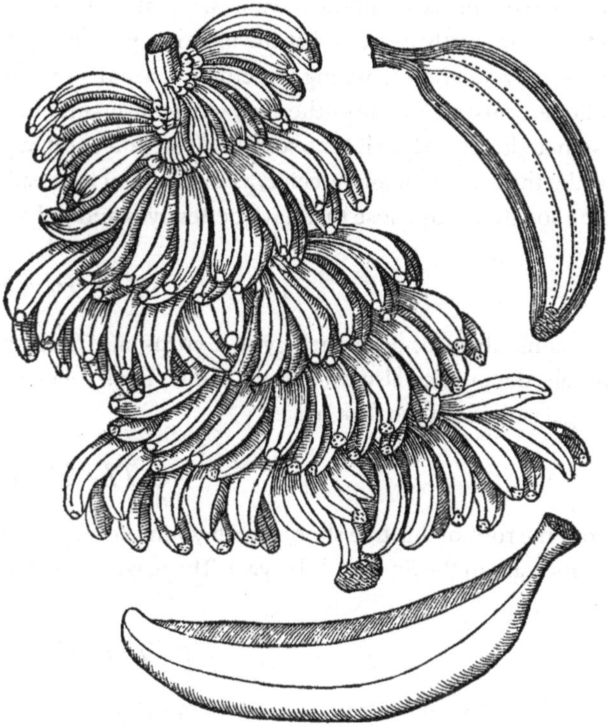

musae fructus exactior icon.
An exacter figure of the Plantaine fruit.

The Description.

Whether this plant may be reckoned for a tree properly or for a herby plant, it is disputable, considering the soft and herby substance whereof it is made; that is to say, when it has attained to the height of six or seven cubits, and of the bigness of a man's thigh, notwithstanding it may be cut down with one stroke of a sword, or two or three cuts with a knife, even with as much ease as the root of a Radish or Carrot of the like bigness.

From a thick fat thready root rise immediately diverse great leaves, of the length of three cubits and a half, sometimes more, according to the soil where it grows, and of a cubit and more broad, of bigness sufficient to wrap a child in of two years old; in shape like those of Mandrake, of an overworn green colour, having a broad rib running through the middle thereof. Which leaves, whether by reason of the extreme hot scorching sun or of their own nature, in September are so dry and withered that there is nothing thereof left or to be seen but only the middle rib. From the midst of these leaves rises up a thick trunk, whereon do grow the like leaves which the people do cut off, as also those next the ground, by means whereof it rises up to the height of a tree, which otherwise would remain a low and base plant. This manner of cutting they use from time to time, until it come to a certain height above the reach of the Elephant, which greedily seeks after the fruit. In the middest of the top, among the leaves, comes forth a soft and fungous stump whereon do grow diverse Apples, in form like a small Cucumber and of the same bigness, covered with a thin rind like that of a Fig, of a yellow colour when they be ripe. The pulp or substance of the meat is like that of the Pompion, without either seeds, stones, or kernels, in taste not greatly perceived at the first, but presently after it pleases and entices a man to eat liberally thereof, by a certain enticing sweetness it yields. In which fruit, if it be cut according to the length, oblique, transverse, or any other way whatsoever, may be seen the shape and form of a cross, with a man fastened thereto. My self have seen the fruit and cut it in pieces, which was brought me from Aleppo in pickle; the cross I might perceive, as the form of a spread-eagle in the root of Fern; but the man I leave to be sought for by those that have better eyes and judgement than my self.

The Place.

This admirable tree grows in Egypt, Cyprus, and Syria, near unto a chief city there called Alep which we call Aleppo; and also by Tripolis not far from thence. It grows also in Canara, Decan, Guzarate, and Bengala, places of the East Indies.

The Time.

From the root of this tree shoots forth young springs or shoots, which the people take up and plant for the increase in the Spring of the year. The leaves wither away in September, as is above said.

The Names.

It is called *musa* by such as travel to Aleppo; by the Arabians, *musa maum;* in Syria, *mose.* The Grecians and Christians which inhabit Syria, and the Jewes also, suppose it to be that tree of whose fruit Adam did taste; which others think to be a ridiculous fable. It is called in the East Indies (as at Malavar where it also grows), *palan*; in Malayo, *pican*; and in that part of Africa which we call Ginny, *bananas*. In English, Adam's Apple tree.

The Temperature and Vertues.

Dioscorides and Serapio judge that it heats in the end of the first degree and moistens in the end of the same. The fruit hereof yields but little nourishment. It is good for the heat of the breast, lungs, and bladder. It stops the liver, and hurts the stomach if too much of it be eaten, and procures looseness in the belly; whereupon it is requisite for such as are of a cold constitution, in the eating thereof to put unto it a little Ginger or other spice. It is also good for the reines or kidneys, and to provoke urine. It nourishes the child in the mother's womb, and stirs to generation.

Of the Alder tree.

The Description.

The Alder tree or Aller, is a great high tree having many brittle branches. The bark is of a brown colour; the wood or timber is not hard, and yet it will last and endure very long under the water, yea longer than any other timber whatsoever. Wherefore in the fenny and soft marish grounds they do use to make piles and posts thereof, for the strengthening of the walls and such like. This timber does also serve very well to make troughs to convey water in stead of pipes of lead. The leaves of this tree are in shape somewhat like the Hasell, but they are blacker and more wrinkled, very clammy to handle, as though they were sprinkled with honey. The blossom or flowers are like the aglets of the Birch tree; which being faded, there follows a scaly fruit closely growing together, as big as a Pigeon's egg, which toward Autumn does open and the seed falls out and is lost.

alnus.
The Alder tree.

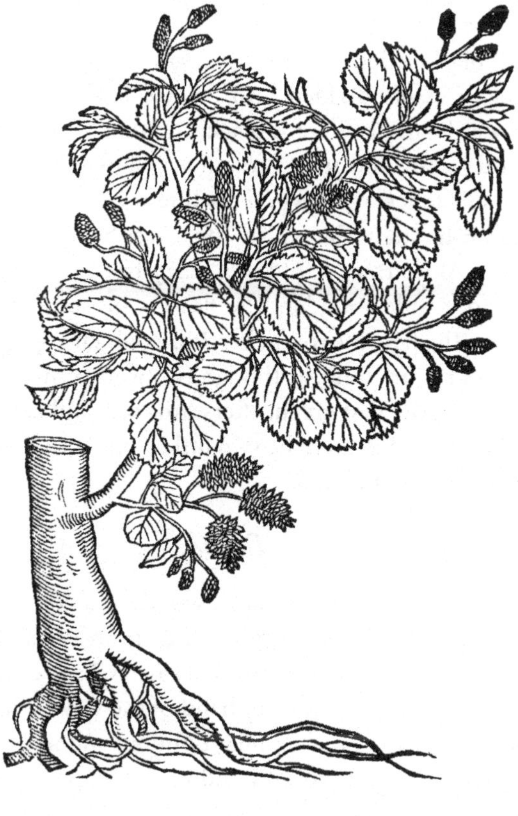

alnus hirsuta.
Rough leaved Alder.

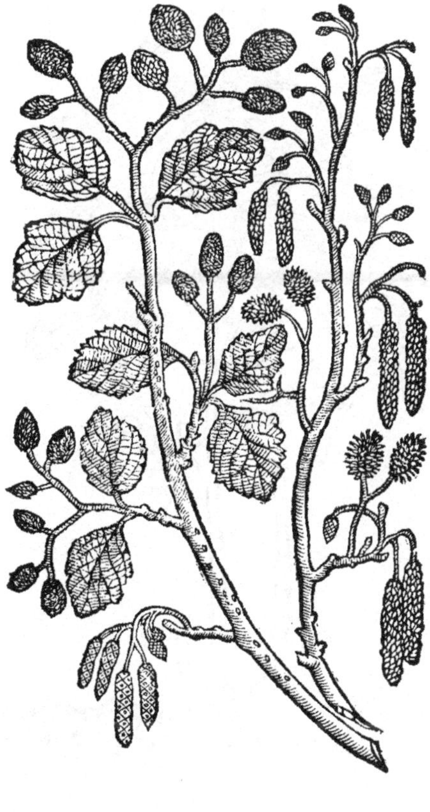

The Place and Time.

The Alder tree delights to grow in low and moist waterish places. The Alder tree brings forth new leaves in April, the fruit whereof is ripe in September.

The Names.

This tree is called in Latin, *alnus;* it is called in High Dutch, *erlenbaum;* in Italian, *alno;* in French, *aulne;* in English, Alder and Aller.

The Temperature and Vertues.

The leaves and bark of the Alder tree are cold, dry and astringent. The leaves of Alder are much used against hot swellings, ulcers, and all inward inflammations, especially of the almonds and kernels of the throat. The bark is much used of poor country dyers, for the dying of coarse cloth, caps, hose, and such like into a black colour, whereunto it serves very well.

alnus nigra, siue frangula.
The blacke Aller tree.

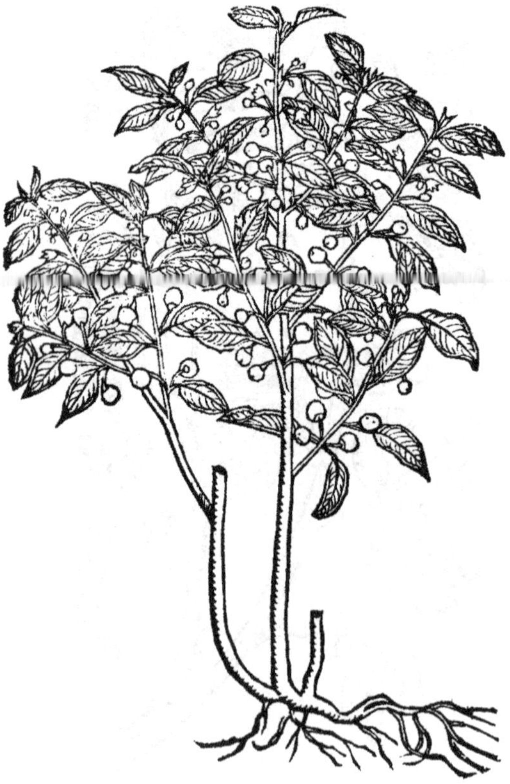

Of the blacke Aller tree.

The Description.

The black Aller tree brings forth from the root straight stalks divided into diverse branches, the outward bark whereof is black, and that next to the wood yellow, and gives a colour as yellow as Saffron. The substance of the wood is white and brittle, with a reddish pith in the midst. The leaves be like those of the Alder tree, or of the Cherry tree, yet blacker and a little rounder. The flowers be somewhat white. The fruit are round berries, in which appear a certain rift or chink, as though two were joined together; at the first green, afterwards red, and last of all black; in this there be two little stones. The root runs along in the earth.

The Place.

The Aller tree grows in moist woods and copses; I found plenty of it in a wood a mile from Islington, in the way from thence toward a small village called Harnsey, lying on the right hand of the way; and in the woods at Hampsted near London, and in most woods in the parts about London.

The Time.

The leaves and flowers appear in the beginning of the Spring, and the berries in Autumn.

The Names.

This shrub is called *alnus nigra,* or *blacke Alder*; and by others, *frangula.* Petrus Crescentius names it *avornus.* In Low Dutch, *sparkenhout* and oftentimes, *pijlhout,* because boys make for themselves arrows hereof. In High Dutch, *faulbaum.* It is called in English, black Aller; and of diverse, Butcher's Pricke tree.

The Temperature and Vertues.

The inner bark of the black Aller tree is of a purging and dry quality.

The inner bark hereof is used of diverse country men, who drink the infusion thereof when they would be purged. It purges thick phlegmaticke humors, and also cholericke, and not only by the stool but also many times by vomit, not without great trouble and pain to the stomacke. It is therefore a medicine more fit for clownes than for civil people; and rather for those that feed grossly than for dainty people.

There be others who affirm that the dried bark is more gentle, and causes lesser pain; for the green bark (say they) which is not yet dried contains in it a certain superfluous moisture which causes gripings and vomitings, and troubles the stomacke.

The same bark being boiled in wine or vinegar makes a lotion for the tooth ache, and is commended against scabs and filthiness of the skin.

The leaves are reported to be good fodder for cattle, especially for kine, and to cause them to yield good store of milk.

Of the Almond tree.

The Description.

The Almond tree is like to the Peach tree, yet it is higher, bigger, and of longer continuance. The leaves be very long, sharp pointed, and snipt about the edges like those of the Peach tree. The flowers be alike; the fruit is also like a Peach, having on one side a cleft, with a soft skin without, and covered with a thin cotton; but under this there is none or very little pulp, which is hard like a gristle and not eaten. The nut or stone within is longer than that of the Peach, not so rugged, but smooth; in which is contained the kernel, in taste sweet and many times bitter. The root of the tree grows deep. The gum which soaks out hereof is like that of the Peach tree.

The Place and Time.

The natural place of the Almond is in the hot regions, yet we have them in our London gardens and orchards in great plenty. The Almond flowers betimes with the Peach. The fruit is ripe in August.

amygdalus.
The Almond tree.

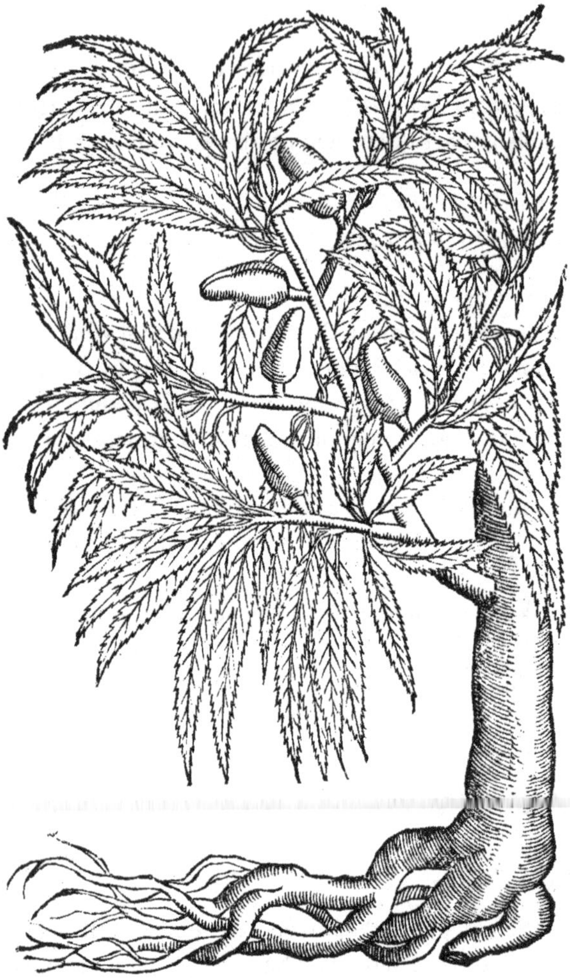

The Names.

The tree is called in Latin, *amygdalus;* in French, *amandier;* in English, Almond tree. The fruit is called in Latin, *amygdalum;* in shops, *amygdala;* in High Dutch, *mandel;* in Low Dutch, *amandelen;* in Italian, *mandole;* in Spanish, *almendras;* in French, *amandes;* in English, Almond.

The Temperature and Vertues.

Sweet Almonds when they be dry be moderately hot, but the bitter ones are hot and dry in the second degree. There is in both of them a certain fat and oily substance, which is drawn out by pressing. Sweet Almonds, being new gathered, are pleasant to the taste; they yield some kind of nourishment, but the same gross and earthy, and grosser than those that be dry and not as yet withered. These do likewise slowly descend, especially being eaten without their skins, for even as the husks or branny parts of corne do serve to drive down the gross excrements of the belly, so do likewise the skins or husks of the Almonds. Therefore, those that be blanched do so slowly descend, as that they do withal bind the belly; whereupon they are given with good success to those that have the laske or the bloody flix.

There is drawn out of sweet Almonds, with liquor added, a white juice like milk, which over and besides that it nourishes, and is good for those that are troubled with the laske and bloody flix, it is profitable for those that have the pleurisy and spit up filthy matter.

Almonds taken before meat do stop the belly and nourish but little; notwithstanding many excellent meats and medicines are therewith made for sundry griefs, yea very delicate and wholesome meats, as Almond butter, cream of Almonds, marchpane, and such like, which dry and stay the belly more than the extracted juice or milk.

The oile which is newly pressed out of the sweet Almonds is a mitigator of pain and all manner of aches. It is good for women that are newly delivered, for it quickly removes the throwes which remain after their delivery.

The oile of Almonds makes smooth the hands and face of delicate persons, and cleanses the skin from all spots, pimples, and lentils. With honey, they are laid upon the bitings of mad dogs. Being applied to the temples with vinegar or oile of Roses, they take away the head ache.

Of the Apple tree.

The Description.

malus carbonaria.

The Pome Water tree.

The Apple tree has a body or trunk commonly of a mean bigness, not very high, having long arms or branches, and the same disordered. The bark is somewhat plain and not very rugged. The leaves be also broad, more long than round, and finely nicked in the edges. The flowers are whitish, tending unto a blush colour.

The fruit or Apples do differ in greatness, form, colour, and taste; some covered with a red skin, others yellow or green, varying infinitely according to the soil and climate, some very great, some little, and many of a middle sort. Some are sweet of taste, or something sour, most be of a middle taste between sweet and sour, the which to distinguish I think it impossible. Notwithstanding, I hear of one that intends to write a peculiar volume of Apples, and the use of them, yet when he has done what he can do, he has done nothing touching their several kinds to distinguish them. This that has been said shall suffice for our History.

The Place.

The tame and grafted Apple trees are planted and set in gardens and orchards made for that purpose. They delight to grow in good and fertile grounds. Kent does abound with Apples of most sorts; but I have seen in the pastures and hedgerows about the grounds of a worshipful Gentleman, dwelling two miles from Hereford, called Master Roger Bodnome, so many trees of all sorts that the servants drink for the most part no other drink than that which is made of Apples. The quantity is such, that by the report of the Gentleman himself, the Parson has for tithe many hogsheads of cider. The hogs are fed with the fallings of them, which are so many that they make choice of those Apples they do eat, who will not taste of any but of the best. An example doubtless to be followed of Gentlemen that have land and living; but envy says the poor will break down our hedges and we shall have the least part of the fruit; but forward in the name of God: set, plant and nourish up trees in every corner of your grounds, the labour is small, the cost is nothing, the commodity is great, your selves shall have plenty, the poor shall have somewhat in time of want to relieve their necessity, and God shall reward your good minds and diligence.

The Time.

They bloom about the end of April, or in the beginning of May. The forward Apples be ripe about the Calends of July, others in September.

The Names.

The Apple tree is called in Latin, *malus* and *pomus;* in High Dutch, *opffelbaum;* in Low Dutch, *appelboom;* in French, *pommier;* in English, Apple-tree. The Latins name the fruit, *malum* or *pomum;* in High Dutch, *opfell;* in Low Dutch, *appel;* in French and Spanish, *mansanas;* in English, Apple.

The Temperature.

All Apples be of temperature cold and moist, and have joined with them a certain excremental or superfluous moisture. But as they be not all of like coldness, so neither have they like quality of superfluous moisture. They are soonest rotten that have greatest store of moisture, and they may be longer kept in which there is less store; for the abundance of excremental moisture is the cause why they rot.

Sweet Apples are not so cold and moist, which being roasted or boiled or otherwise kept, retain or keep the soundness of their pulp. They yield more nourishment, and not so moist a nourishment, as do the other Apples, and do not so easily pass through the belly. Sour Apples are colder and also moister. The substance or pulp of these when they be boiled does run abroad and retains not his soundness; yet they yield a lesser nourishment, and the same raw and cold. They do easily and speedily pass through the belly and therefore they do mollify the belly, especially being taken before meat.

The Vertues.

Roasted Apples are always better than the raw, the harm whereof is both mended by the fire, and may also be corrected by adding unto them seeds or spices. The juice of Apples which be sweet and of a middle taste, is mixed in compositions of diverse medicines, and also for the tempering of melancholy humours, and likewise to mend the qualities of medicines that are dry. There is likewise an ointment with the pulp of Apples and swine's grease and Rose water, which is used to beautify the face and to take away the roughness of the skin, which is called in shops *pomatum,* of the Apples whereof it is made.

The pulp of the roasted Apples, in number four or five according to the greatness of the Apples, mixed in a wine quart of fair water, laboured together until it come to be as Apples and ale, which we call Lambes Wooll, and the whole quart drunk last at night within the space of an hour, does in one night cure those that pisse by drops with great anguish and dolour, the strangerie, and all other diseases proceeding of the difficulty of making water; but in twice taking it, it never fails in any. Oftentimes there happens with the foresaid diseases the gonorrhæa, or running of the reines, which it likewise heals in those persons, but not generally in all, which my self have often proved and gained thereby both crowns and credit.

Apples cut in pieces and distilled with a quantity of Camphere and butter-milk, take away the marks and scars gotten by the small-pockes, being washed therewith when they grow into their state and ripeness; provided that you give unto the patient a little milk and Saffron to drink, to expel to the extreme parts that venom which may lie hid, and as yet not seen.

Of the Aprecocke or Abrecocke tree.

The Description.

This tree is greater than the Peach tree, and has a bigger body. It lasts longer, especially if it be grafted or inoculated. The leaves hereof are broad, sharp pointed, and nicked in the edges. The flowers are somewhat white. The fruit is round like a Peach, yellow within and without, in which does lie a brown stone, nothing rough at all as is that of the Peach, shorter also and lesser, in which is included a sweet kernel.

The Place and Time.

These trees do grow in my garden, and nowadays in many other Gentlemen's gardens throughout all England. They flower and flourish in April, and their fruit is ripe in July.

The Names.

This tree is called in Latin, *malus armeniaca;* in English, Abrecocke tree and Aprecocke tree. The fruit is named *malum armeniacum;* in Italian, *armeniache;* in French, *abricoz;* in English, Abrecocke, and of some, Aprecocke and Aprecox.

armeniaca malus maior.

The greater Aprecocke tree.

The Temperature and Vertues.

Aprecockes are cold and moist in the second degree, but yet not so moist as Peaches, for which cause they do not so soon or easily putrefy, and they are also more wholesome for the stomach and pleasant to the taste; yet do they likewise putrefy, and yield little nourishment. Being taken after meat, they corrupt and putrefy in the stomach; being first eaten before other meat, they easily descend and cause the other meats to pass down the sooner, like as also the Peaches do. The vertues of the leaves of this tree are not as yet found out.

Of the Ash tree.

fraxinus.
The Ash tree.

The Description.

The Ash is a high and tall tree. It rises up with a straight body, now and then of no small bigness, now and then of a middle size, and is covered with a smooth bark. The wood is white, smooth, hard, and somewhat rough-grained. The tender branches hereof, and such as be new grown-up, are set with certain joints and have within a white and spongy pith; but the old boughs are woody throughout, and be without either joints or much pith. The leaves are long and winged, consisting of many standing by couples, one right against another upon one rib or stalk, the uppermost of all excepted, which stands alone; of which every particular one is long, broad, like to a Bay leaf, but softer and of a lighter green, without any sweet smell, and nicked round about the edges. Out of the younger sort of the boughs grow forth hanging together many long narrow and flat cods, as it were like almost to diverse birds' tongues, where the seed is perfected, which is of a bitter taste. The roots be many, and grow deep in the ground.

The Place.

The Ash does better prosper in moist places, as about the borders of meadows and river sides, than in dry grounds.

The Time.

The leaves and keyes come forth in April and May, yet is not the seed ripe before the fall of the leaf.

The Names.

This tree is called in Latin, *fraxinus;* in High Dutch, *eschernbaum;* in Italian, *frassino;* in French, *fresne;* in Spanish, *fresno, fraxino,* and *freixo;* in English, Ash tree. The fruit like unto cods is called of the Apothecaries, *lingua avis* and *lingua passerina.* It is termed in English, Ash Keyes and of some, Kite-keyes.

The Temperature and Vertues.

The leaves and bark of the Ash tree are dry and moderately hot. The seed is hot and dry in the second degree. The juice of the leaves, or the leaves themselves being applied or taken with wine, cure the bitings of vipers, as Dioscorides says. The leaves of this tree are of so great vertue against serpents, as that they dare not so much as touch the morning and evening shadows of the tree, but shun them afar off, as Pliny reports. He also affirms that the serpent, being penned in with the boughs laid round about, will sooner run into the fire, if any be there, than come near the boughs of the Ash. It is a wonderful courtesy in nature that the Ash should flower before these serpents appear, and not cast his leaves before they be gone again.

Both of them, that is to say the leaves and the bark, are reported to stop the belly; and being boiled with vinegar and water, do stay vomiting if they be laid upon the stomach. The seed or kite-keys of the Ash tree provoke urine, increase natural seed, and stir up bodily lust, especially being powdered with nutmegs and drunk. The lee which is made with the ashes of the bark, cures the white scurf and such other like roughness of the skin, as Pliny testifies.

Of the wilde Ash, otherwise called Quicke-Beame or Quicken tree.

The Description.

The wilde Ashe or Quicken Tree *Pena* sets forth for the wild Service. This tree grows seldom or never to the stature and height of the Ash tree, notwithstanding it grows to the bigness of a large tree. The leaves be great and long, and scarcely be discerned from the leaves of the Service tree. The flowers be white and sweet of smell, and grow in tufts which do turn into round berries, green at the first but when they be ripe, of a deep red colour and of an unpleasant taste. The branches are as full of juice as the Osiar, which is the cause that boys do make pipes of the bark thereof, as they do of Willows.

The Place.

The wilde Ashe or Quicken Tree grows on high mountains, and in thick high woods in most places of England, especially about Namptwich in Cheshire, in the Weilds of Kent, in Sussex and diverse other places.

sorbus sylvestris, siue fraxinus bubula.
The Quicken tree, wilde Aſh,
or wilde Service tree.

The Time.

The wild Ash flowers in May, and the berries are ripe in September.

The Names.

The Latines call this tree *Ornus,* and oftentimes *sylvestris fraxinus,* or wilde Ash, and it is also *fraxini species,* or a kinde of Ash; for the Grecians (as not only Pliny writes but also Theophrastus) have made two kinds of Ash, the one high and tall, the other lower. The high and tall one is *fraxinis vulgaris,* or the common Ash; the lower *ornus,* which also is named *montana fraxinus,* mountaine Ash, as the other is named field Ash, which is also named *magna fraxinus* or great Ash.

This *ornus* or great Ash is named in High Dutch, *malbaum;* in Low Dutch, *haveresschen,* or *quereschen,* of diverse, *qualster;* in French, *fresne sauvage;* in English, Wilde Ash, Quicken tree, Quick-beame tree, and Wicken tree. Matthiolus makes this to be *serbus sylvestris* or wilde Service tree.

The Temperature and Vertues.

Touching the faculties of the leaves, bark, or berries, as there is nothing found among the old, so is there nothing noted among the later writers, but Pliny seems to make this wilde Ash like in faculties to the common Ash.

The leaves of the wilde Ash tree boiled in wine are good against the pain of the sides and the stopping of the liver, and assuage the bellies of those that have the tympanie and dropsie.

Benedictus Curtius Symphoryanus is deceived in the history of *ornus,* when he thinks out of Virgil's Georgicks, that *ornus* has the flower of the Peare tree; for out of Virgil's verses no such thing at all can be gathered. For he intreateth not of the forms of the trees, but of the graffing of diverse into others, unlike and differing in nature; as of the graffing of the Nut tree into the Strawberry tree; the Apple into the Plane tree, the Beech into the Chestnut tree; the Peare into the wilde Ash or Quick-beame tree; the Oke into the Elme tree; and in this respect he writes that the Plane tree brings forth an Apple; the Beech tree a Chestnut; the wilde Ash tree brings forth the white flower of the Peare tree, as is most manifest out of Virgil's own words, after this manner, in the second book of his Georgicks:

The Tree Strawb'ry on Walnuts stocke doth grow,
And barren Planes faire Apples oft have borne;

Chestnuts, Beech-Mast; the Quicken tree doth shew
The Peares white floure; and twine oft times th' Acorn
Have gathered under Elmes.

Of the Balsam tree.

The Description.

The Balsam tree grows to the height and bigness of the Pomegranate tree, garnished with very many branches whereon do grow leaves like those of Rue, but of colour whiter, always growing green; amongst which come forth flowers, whereof we have no certainty; after which comes forth fruit like that of the Turpentine tree, which in shops is called *Carpobalsamum,* of a pleasant smell; but the liquor which flows from the wounded tree is much sweeter; which liquor of some is called *Opobalsamum.*

balſamum alpini cum carpobalſamo.
The Balſam tree with the fruit.

The Place and Time.

These trees grow in diverse parts of the world, some in Ægypt and most of those countries adjacent, and in the East and West Indies, as travellers in those parts report. For the most part, they keep green Winter and Summer.

The Names.

Balsam is called in Latin, *balsamum;* of the Arabians, *balseni, balesina* and *belsan;* in Italian, *balsamo;* in French, *baume.* The liquor that flows out of the tree when it is wounded is called *opobalsamum;* the fruit, *carpobalsamum;* and the liquor which naturally flows from the tree in Ægypt, *balsamum.*

The Temperature and Vertues.

Balsam is hot and dry in the second degree, with astriction. Natural Balsam taken in a morning fasting, with a little Rose water or wine, to the quantity of five or six drops, helps those that be asthmaticke or short of wind. It prevails against the pains of the bladder and stomach, and comforts the same mightily. It also amends a stinking breath and takes away the shaking fits of the quotidian ague, if it be taken two or three times. It helps consumptions, cleanses the barren womb, especially being anointed upon a pessary or mother suppository, and used. It also refreshes the brain and comforts the parts adjoining; it helps the palsy, convulsions, and all griefs of the sinews, if they be anointed therewith. The marvellous effects that it works in new and green wounds were here too long to set down and also superfluous, considering the skilful Surgeon, whom it most concerns, does know the use thereof; and as for the beggarly Quacksalvers, Runnagates and knavish Mountibanks, we are not willing to instruct them in things so far above their reach, capacity and worthiness.

Of a kinde of Balme, or Balsame Tree.

The Description.

This tree which the people of the Indies do call *molli*, grows to the bigness of a great tree, having a trunk or body of a dark green colour, sprinkled over with many ash coloured spots. The branches are many,

molli, siue molly clusij, & lobelij.
The Balsame tree of Clusius and Lobels defcription.

molle arboris adultae ramus.
A branch of the old tree of Molle.

and of very great beauty; whereupon do grow leaves not unlike to those of the Ash tree, consisting of many small leaves set upon a middle rib, growing narrower ever towards the point, every particular one jagged on the sides like the teeth of a saw; which being plucked from the stem, yieldeth forth a milky juice, tough and clammy, savouring like the bruised leaves of Fennell, and as it seems in taste somewhat astringent. The flowers grow in clusters upon the twiggie branches, like those of the Vine a little before the grapes be formed. After follows the fruit or berries, somewhat greater than Pepper cornes, of an oilie substance, green at the first, and of a dark reddish colour when they be ripe.

The Place.

This tree, sayeth a learned Physician called Ioh. Fragosus, grows in the King of Spaine his garden at Madryll, which was the first that ever he did see; since which time, Iohn Ferdinand, secretary unto the foresaid king, did shew unto the said Fragosus in his own garden a tree so large and of such beauty, that he was never satisfied with looking on it, and meditating upon the vertues thereof. Which words I have received from the hands of a famous learned man, called Mr. Lancelot Browne, Dr. in Physicke, and Physician to the Queenes Majestie, at the impression hereof, faithfully translated out of the Spanish tongue, without adding or taking any thing away.

They grow plentifully in the vales and low grounds of Peru, as all affirm that have travelled to the West Indies, as also those that have described the singularities thereof. My self with diverse others, as namely Mr. Nicholas Lete, a worshipful Merchant of the City of London; and also a most skilfull Apothecary, Mr. James Garret, who have received seeds hereof from the right Honorable the Lord Hunsdon, Lord High Chamberlaine of England, worthy of triple honour for his care in getting, as also for his curious keeping of rare and strange things brought from the farthest parts of the world. Which seeds we have sown in our gardens, where they have brought forth plants of a foot high; and also their beautiful leaves. Notwithstanding our care, diligence, and industry, they have perished at the first approach of winter, as not being able by reason of their tenderness to endure the cold of our Winter blasts.

The Time.

As touching the time of his flourishing and bringing his fruit to maturity, we have as yet no certain knowledge, but is thought to be green both winter and summer.

The Names.

This most notable tree is called by the Indian name *molle;* of some, *molly,* and *muelle,* taken from his tender softness, as some have deemed. It may be called the Fennell tree, or one of the Balme, or Balsam trees.

The Temperature.

This tree is thought to be of an astringent or binding quality, whereby it appears besides the hot temperature it has, to be compounded of diverse other faculties.

The Vertues.

The Indians use to seeth the fruit or berries hereof in water, and by a special skill they have in the boiling, to make a most wholesome wine or drink, as also a kind of vinegar, and sometimes honey, which are very strange effects, these three things being so contrary in taste.

The leaves boiled, and the decoction drunke, helps them of any disease proceeding of a cold cause.

The gum which issues from the tree, being white like unto Manna, dissolved in milk, takes away the web of the eyes, and clears the sight being wiped over with it.

The bark of this tree boiled, and the legs that be swollen and full of pain, bathed and washed with the decoction diverse times, takes away both infirmities in short space.

This tree is of such estimation among the Indians, that they worship it as a god, according unto their savage rites and ceremonies.

Of the Bay or Laurell tree.

The Description.

The Bay or Laurell tree comes often times to the height of a tree of mean bigness. It is full of boughs, covered with a green bark, and the leaves thereof are long, broad, hard, of colour green, sweetly smelling and in taste somewhat bitter. The flowers along the boughs and leaves are of a green colour. The berries are more long than round, and be covered with a black rind or pill. The kernel within is cloven into two parts, like that of the Peach and Almond and other such, of a brown yellowish colour, sweet of smell, in taste somewhat bitter, with a little sharp or biting quality.

Laurus.
The Bay tree.

The Place and Time.

The Bay tree grows naturally in Spaine and such hot regions. We plant and set it in gardens, defending it from cold at the beginning of March especially. I have not seen any one tree thereof growing in Denmark, Sweuia, Poland, Livonia, or Russia, or in any of those cold countries where I have travelled.

The Bay tree grows green Winter and Summer. It flowers in the Spring, and the black fruit is ripe in October.

The Names.

This tree is called in Latin, *laurus;* in Italian, *lauro;* in High Dutch, *looetbeerbaum;* in Low Dutch, *laurus boome;* in French, *laurier;* in Spanish, *laurel* and *lorel;* in English, Laurell or Bay tree. The poets fain that it took his name of Daphne, Lado his daughter, with whom Apollo fell in love.

The Temperature and Vertues.

The berries and leaves of the Bay tree, says Galen, are hot and very dry, and yet the berries more than the leaves. The bark is not biting and hot, but more bitter, and it has also a certain astrictive or binding quality. Bay berries with honey or Cute, are good in a licking medicine, says Dioscorides, against the pthisicke or consumption of the lungs, difficulty of breathing, and all kinds of fluxes or rheums about the chest. Bay berries are put into Mithridate, Treacle, and such like medicines that are made to refresh such people as are grown sluggish and dull by means of taking opiate medicines, or such as have any venomous or poisoned quality in them.

The oile pressed out of these, or drawn forth by decoction, does in short time take away scabs and such like filth of the skin. It cures them that are beaten black and blew, and that be bruised by squats and falls. It removes black and blew spots, and congealed blood, and digests and wastes away the humours gathered about the grieved part. It is reported that common drunkards were accustomed to eat in the morning fasting two leaves thereof against drunkenness. It helps the dropsy and the jaundice, and procures unto women their desired sickness.

Of the Wilde Bay tree.

The Description.

Laurus tinus, or the wilde Bay tree, grows like a shrub or hedge both, having many tough and pliant branches, set full of leaves very like to the Bay leaves, but smaller and more crumpled, of a deep and shining green colour. Among which come forth tufts of whitish flowers, turning at the edges into a light purple; after which follow small berries of a blew colour, containing a few graines or seeds like the stones or seeds of grapes. The leaves and all the parts of the plant are altogether without smell or savour.

Tinus lusitanica, or the Portingale wilde Bay tree, grows very like to *cornus fœmina,* or the Dog-berry tree, but the branches be thicker and more stiff, covered with a reddish bark mixed with green. The leaves are like the former, but larger, having many sinews or veins running through the same like as in the leaves of Sage. The flowers hereof grow in tufts like the precedent, but they are of colour more declining to purple. The small branches are likewise of a purple colour. The leaves have no smell at all, either good or bad. The berries are smaller than the former, of a blew colour declining to blacknesse.

laurus tinus.
The wilde Bay tree.

laurus tinus lusitanica.
The Portingale wilde Bay tree.

The Place.

The wilde Bay grows plentifully in every field of Italy, Spain, and other regions, which differ according to the nature and situation of those countries. They grow in my garden and prosper very well.

The wilde Laurell is ever greene, and may oftentimes be seen most part of the winter, and the beginning of the spring, with the flowers and ripe berries growing both at one season.

The Names.

It is called in Latine, *tinus,* and *laurus sylvestris.* Cato names it *laurus sylvatica.* In Italian, *lauro sylvatico;* in Spanish, *vua de perro,* otherwise *follado;* and of diverse, *durillo;* in English, wilde Bay.

The Temperature and Vertues.

Pliny nor any other of the Ancients have touched the faculties of this wilde Bay, neither have we any understanding thereof by the later writers, or by our own experience.

Of the Beade tree.

The Description.

This tree was called *zizypha candida* by the Herbarists of Montpelier; and by the Venetians and Italians, *sycomorus,* but untruly; the Portugals have termed it *arbor paradizo;* all which and each whereof have erred together, both in respect of the fruit and of the whole tree. Some have called it *zizypha,* though in facultie it is nothing like; for the taste of this fruit is very unpleasant, virulent, and bitter.

But deciding all controversies, this is the tree which Avisen calls *azederach,* which is very great, charged with many large arms that are garnished with twiggie branches, set full of great leaves consisting of sundry small leaves, one growing right opposite to another like the leaves of the Ash tree or Wicken tree, but more deeply cut about the edges like the teeth of a saw. Among which come the flowers, consisting of five small blew leaves laid abroad in manner of a starre; from the middest whereof grows forth a small hollow cup resembling a Chalice. After which succeeds the fruit, covered with a brownish yellow shell, very like unto the fruit of Iuiubes (whereof Dodonaeus in his last edition makes it a kind) of a rank, bitter, and unpleasant taste, with a six cornered stone within, which being drawn on a string, serves to make Beades of, for want of other things.

Zizyphus cappadocica grows not so great as the former, but is of a mean stature, and full of boughs. The bark is smooth and even, and that which grows upon the trunk and great boughs is of a shining scarlet colour. Out of these great arms or boughs grow slender twigs, white and soft, which are set full of whitish leaves, but more white on the contrary or back part, and are like to the leaves of Willow, but narrower and whiter. Amongst these leaves come forth small hollow yellowish flowers, growing at the joints of the branches, most commonly three together and of a pleasant savour, with some few threads or chives in the middle thereof. After which succeeds the fruit, of the bigness and fashion of the smallest Olive, white both within and without, wherein is contained a small stone which yields a kernel of a pleasant taste, and very sweet.

The Place.

Matthiolus writes that *zizyphus candida* is found in the cloisters of many monasteries in Italie; Lobel says that it grows in many places in Venice and Narbon; and it is wont now of late to be planted and cherished in the goodliest orchards of all the low Countries.

Zizyphus cappadocica grows likewise in many places of Italy, and specially in Spaine. It is also cherished in gardens both in Germany and in the low Countries. It grows also here in the garden of M^r^. John Parkinson.

The Time.

These trees flower in June in Italy and Spain; their fruit is ripe in September, but in Germany and the low Countries no fruit follow the flowers.

zizypha candida.
The Beade tree.

zizypha cappadocica.
The Beade tree of Cappadocia.

The Names.

Zizyphus candida Avicen calls *azederach,* or as diverse read it, *azederaeth.* The later writers are far deceived in taking it to be the Sycomore tree; and they as much that would have it to be the Lote or Nettle tree. It may be named in English, Bead tree, for the cause before alleged.

The other is *altera species zizyphi,* or the second kind of Iuiube tree, which Columella in his ninth book and fourth chap. does call *zizyphus alba,* or white Iuiube tree, for difference from the other that is surnamed *rutila,* or glittering red. Pliny calls this *zizyphus cappadocica.*

The Temperature.

Avicen writing of *azadaraeth,* says that the flowers thereof be hot in the third degree, and dry in the end of the first.

Zizyphus Cappadodica is cold and dry of complexion.

The Vertues.

The flowers of *zizyphus,* or *azadaraeth,* open the obstructions of the braine.

The distilled water thereof kills nits and lice, preserves the hair of the head from falling, especially being mixed with white wine, and the head bathed with it.

The fruit is very hurtful to the chest, and a troublesome enemy to the stomacke; it is dangerous, and peradventure deadly.

Moreover, it is reported that the decoction of the bark and of Fumitorie, with Mirobalans added, is good for agues proceeding of flegme.

The juice of the uppermost leaves with honey is a remedie against poison.

Matthiolus writes that the leaves and wood bring death even unto beasts, and that the poison thereof is resisted by the same remedies that Oleander is.

Zizyphus cappadocica prevails against the diseases aforesaid, but the decoction thereof is very good for those whose water scalds them with the continual issuing thereof, as also for such as have the running of the reines and the exulcerations of the bladder and privy parts.

A looch or licking medicine made of the syrup, is excellent good against spitting of blood proceeding of the distillations of sharp or salt humors.

Of the Beech tree.

The Description.

The Beech is a high tree with boughs spreading oftentimes in manner of a circle, and with a thick body having many arms. The bark is smooth. The timber is white, hard, and very profitable. The leaves be smooth, thin, and broad. The catkins or blowings be also lesser and shorter than those of the Birch tree, and yellow. The fruit or mast is contained in a husk or cup that is prickly and rough bristled, yet not so much as that of the Chestnut; which fruit being taken forth of the shells or urchin husks, be covered with a soft and smooth skin, like in colour and smoothness to the Chestnuts, but they be much lesser and of another form, that is to say, triangled or three cornered. The kernel within is sweet, with a certain astriction or binding quality. The roots be few, and grow not deep, and little lower than under the turf.

The Place.

The Beech tree loves a plain and open country, and grows very plentifully in many forests and desert places of Sussex, Kent, and sundry other counties.

The Time.

The Beech flowers in April and May, and the fruit is ripe in September, at what time the deer do eat the same very greedily, as greatly delighting therein; which has caused foresters and huntsmen to call it Buck-mast.

The Names.

The tree is called in Latin, *fagus;* in High Dutch, *buchbaum* or *buch;* in Low Dutch, *bukenboom;* in Italian, *faggi;* in Spanish, *faia* and *fax;* in French, *hestre;* in English, Beech tree, Beech-mast, and Buck-mast. The fruit is called in Latin, *nuces fagi;* in Low Dutch, *bueken nootkens;* in French, *faine;* in English, Beech-mast.

The Temperature.

The leaves of Beech do cool; the kernel of the nut is somewhat moist.

The Vertues.

The leaves of Beech are very profitably applied unto hot swellings, blisters, and excoriations; and being chewed, they are good for chapped lips and pain of the gums.

The kernels or mast within are reported to ease the pain of the kidneys proceeding of the stone, if they be eaten, and to cause the gravel and sand the easier to come forth. With these, mice and squirrels are greatly delighted, who do mightily increase by feeding thereon.

Swine also be fattened herewith, and certain other beasts; also deer do feed thereon very greedily. They be likewise pleasant to thrushes and pigeons. Petrus Crescentius writes that the ashes of the wood is good to make glass with. The water that is found in the hollowness of Beeches cures the naughty scurf, tetters, and scabs of men, horses, kine, and sheep, if they be washed therewith.

fagus.
The Beech.

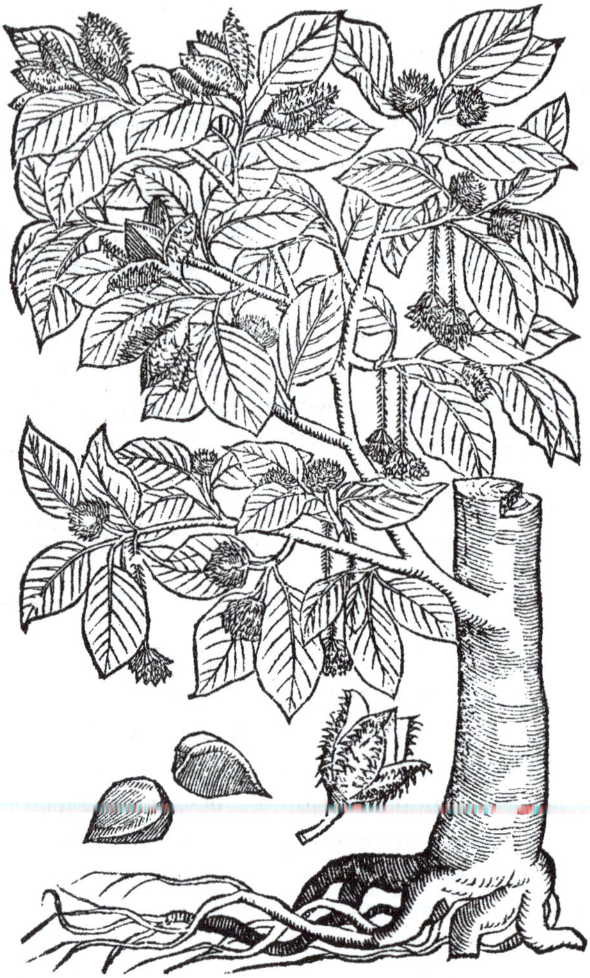

Of the Birch tree.

The Description.

The common Birch tree waxes a great tree, having many boughs beset with many small rods or twigs, very limber and pliant. The bark of the young twigs and branches is plain, smooth, and full of sap, in colour like the Chestnut; but the rind of the body or trunk is hard without, white, rough, and uneven, full of chinks or crevices under which is found another fine bark, plain, smooth, and as thin as paper, which heretofore was used in stead of paper to write upon, before the making of paper was known. In Russia and these cold regions it serves in stead of tiles and slate to cover their houses withal. This tree bears for his flowers certain aglets, like the Hasell tree but smaller, wherein the seed is contained.

The Place and Time.

This common Birch tree grows in woods, fenny grounds, and mountains, in most places of England. The catkins or aglets do first appear, and then the leaves, in April or a little later.

betula.

The Birch tree.

The Names.

The Birch tree is called in Latin, *betula;* diverse also write it with a double ll, *betulla,* as some of Pliny's copies have it. It is called in High Dutch, *birkenbaum;* in Low Dutch, *berckenboom;* in Italian, *betula;* by them of Trent, *bedallo;* in French, *bouleau;* in English, Birch tree.

The Vertues.

Concerning the medicinable use of the Birch tree or his parts, there is nothing extant either in the old or new writers. This tree, says Pliny, *mirabili candore & tenuitate terribilis magistratuum virgis,* for in times past, the Magistrates' rods were made hereof. And in our time also, the Schoolmasters and parents do terrify their children with rods made of Birch.

It serves well to the decking up of houses and banqueting rooms, for places of pleasure, and beautifying of streets in the Crosse or Gang Weeke, and such like.

Of the Box Tree.

The Description.

The great Box is a faire tree, bearing a great body or trunk. The wood or timber is yellow and very hard, and fit for sundry works, having many boughs and hard branches, beset with sundry small hard green leaves, both winter and summer like the Bay tree. The flowers are very little, growing among the leaves, of a green colour, which being faded there succeed small black shining berries, of the bigness of the seeds of Corianders, which are inclosed in round greenish husks, having three feet or legs like a brasse or boiling pot. The root is likewise yellow, and harder than the timber, but of greater beauty and more fit for dagger haftes, boxes, and such like uses, whereto the trunk or body serves, than to make medicines; though foolish empericks and women leeches do minister it against the Apoplexie and such diseases. Turners and Cutlers, if I mistake not the matter, do call this wood Dudgeon, wherewith they make Dudgeon hafted daggers.

There is also a certain other kind hereof, growing low and not above half a yard high, but it spreads all abroad. The branches hereof are many and very slender; the leaves be round, and of a light green.

buxus.

The Box tree.

The Place and Time.

Buxus, or the Box tree, grows upon sundry waste and barren hills in England, and in diverse gardens.

The Box tree grows green winter and summer. It flowers in February and March, and the seed is ripe in September.

The Names.

In Latine, *buxus;* in High Dutch, *buchszbaum;* in Low Dutch, *burboom;* in Italian, *bosso;* in English, Box tree.

The lesser may be called in Latine, *humi buxus,* or *humilis buxus;* in English, dwarf Box, or ground Box, and it is commonly called Dutch Box.

The Temperature and Vertues.

The leaves of the Box tree are hot, dry, and astringent, of an evil and loathsome smell, not used in medicine, but only as I said before in the description.

Of the Carob tree, or Saint Iohns Bread.

The Description.

The Carob tree is one of those that bear cods. It is a tree of middle bigness, very full of boughs. The leaves be long and consist of many set together upon one middle rib, like those of the Ash, but every particular one of them is broader, harder, and rounder. The fruit or long cods in some places are a foot in length, in other places shorter by half, an inch broad, smooth and thick; in which do lie flat and broad seeds. The cods themselves are of a sweet taste and are eaten of diverse, but not before they be gathered and dried; for being as yet green, though ripe, they are unpleasant to be eaten by reason of their ill favoured taste.

The Place.

This grows in Apulia, a province of the Kingdom of Naples, and also in diverse untoiled places in Spaine. It is likewise found in India, and other countries eastward, where the cods are so full of sweet juice as that it is used to preserve Ginger and other fruits. Aristobulus reports how there is a tree in India of no great bigness, which has great cods, ten inches long, full of honey. It is very well known in the coasts of Nicea and Liguria in Italy, as also in the tracts and coasts of the West Indies and Virginia. It grows also in sundry places of Palestine, where there is such plenty of it that it is left unto swine and other wild beasts to feed upon, as our Acorns and Beech mast. Moreover, both young and old feed thereon for pleasure, and some have eaten thereof to supply and help the necessary nourishment of their bodies. This of some is called Saint John's Bread, and thought to be that which is translated Locusts, whereon S. John did feed when he was in the wilderness, besides the wild honey whereof he did also eat, but there is small certainty of this; but most certain that the people of that country do feed upon these cods.

The Time.

The Carob tree brings forth fruit in the beginning of the Spring, which is not ripe till Autumn.

ceratia siliqua, siue ceratonia.
The Carob tree.

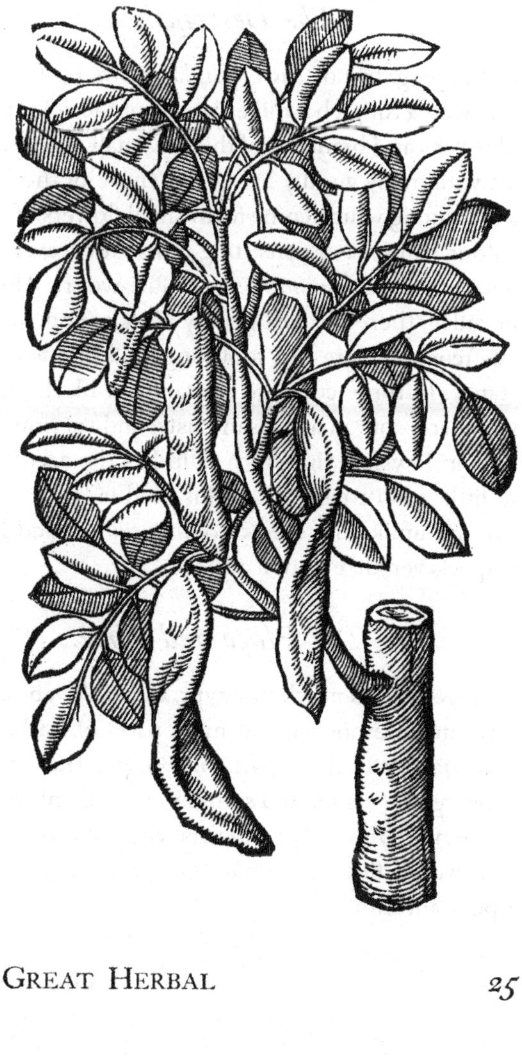

The Names.

The Carob tree is called in Latin, *ceratonia.* In Spanish, *garovo;* in English, Carob tree; and of some, Beane tree and Saint John's Bread. In High Dutch, *s. johansbrot;* neither is it known by any other name in the Low Countries. Some call it in English, Carob.

The Temperature and Vertues.

The Carob tree is dry and astringent, as is also the fruit, and contains in it a certain sweetness, as Galen says. The fruit of the Carob tree, being eaten when it is green, does gently loose the belly; but being dry, it is hard of digestion and stops the belly. It provokes urine, it is good for the stomach and nourishes well, and much better than when it is green and fresh.

Of Cassia Fistula, or Pudding Pipe.

The Description.

Cassia purgatrix or *cassia fistula,* grows up to be a fair tree with a tough bark like leather, of the colour of Box. The arms and branches of this are small and limber, beset with many goodly leaves like those of the Wall-nut tree; among which come forth small flowers of a yellow colour, compact or consisting of six little leaves, like the flowers of *chelidonium minus* or Pile-woort. After these be faded, there succeed goodly black round, long cods, whereof some are two foot long, and of a woody substance. In these cods are contained a black pulp, very sweet and soft, of a pleasant taste, and serving to many uses in Physicke, in which pulp lies the seed couched in little cels or partitions. This seed is flat and brownish, not unlike the seed of *ceratia siliqua,* and in other respects very like unto it also.

cassia fistula.
Pudding Pipe tree.

The Place and Time.

This tree grows much in Egypt, especially about Memphis and Alexandria, and most parts of Barbarie, and is a stranger in these parts of Europe. The Cassia tree grows green winter and summer. It sheds his old leaves when new are come, by means whereof it is never void of leaves. It flowers early in the spring, and the fruit is ripe in autumn.

The Names.

This tree was unknown to the old writers, or so little accounted of, as that they have made no mention of it at all. The Arabians were the first that esteemed of it, by reason they knew the use of the pulp which is found in the Pipes; the fruit thereof, saith Actuarius in his fifth book, is like a long pipe, having within it a thick humour or moisture, which is not congealed all alike through the pipe, but is separated and divided with many partitions, being thin woody skins. The Apothecaries call it *casia fistula* and *cassia fistula;* it is called in English after the Apothecaries' word, *cassia fistula,* and may also be Englished, Pudding Pipe, because the cod or Pipe is like a pudding. But the old *cassia fistula* is that sweet and odoriferous bark that is rolled together, after the manner of a long and round pipe, now named of the Apothecaries *cassia lignes,* which is a kind of Cinnamon.

The Temperature.

The pulp of this pipe which is chiefly in request, is moist in the later end of the first degree, and little more than temperately hot.

The Vertues.

The pulp of *cassia fistula* extracted with violet water, is a most sweet and pleasant medicine, and may be given without danger to all weak people of what age and sex soever they may be; yea it may be ministered to women with child, for it gently purges cholericke humours and slimie flegme, if it be taken in the weight of an ounce.

Cassia is good for such as be vexed with hot agues, pleurisies, jaundice, or any other inflammation of the liver, being taken as afore is shewed.

Cassia is good for the reines and kidneys, drives forth gravel and the stone, especially if it be mingled with the decoction of Parsley and Fennel roots, and drunk.

It purges and purifies the blood, making it more clean than before, breaking therewith the acrimonie and sharpness of the mixture of blood and choler together.

It dissolves all phlegmons and inflammations of the breast, lungs, and the rough artery called *trachea arteria,* easing those parts exceeding well.

Cassia abates the vehemence of thirst in agues, or any hot disease whatsoever, especially if it be taken with the juice of *intybum, cichoreum,* or *solanum,* depured according to Art. It abates also the intemperate heat of the reines, if it be received with diureticke simples, or with the decoction of Licorice only, and will not suffer the stone to grow in such persons as do receive and use this medicine.

The best *cassia* for your use is to be taken out of the most full, most heavy, & faire cods or canes, and those which do shine without and are full of soft pulp within. That pulp which is newly taken forth is better than that which is kept in boxes, by what Art soever.

Many singular compounded medicines are made with this *cassia,* which here to recite belongs not to my purpose or history.

Of the Cedar tree.

cedrus libani.

The great Cedar tree of Libanus.

The Description.

The great Cedar is a very big and high tree, not only exceeding all other resinous trees, and those which bear fruit like unto it, but in his tallness and largeness far surmounting all other trees. The body or trunk thereof is commonly of a mighty bigness, insomuch as four men are not able to fathom it, as Theophrastus writes. The bark of the lower part, which proceeds out of the earth, to the first young branches or shoots, is rough or harsh; the rest which is among the boughs, is smooth and glib. The boughs grow forth almost from the bottom and not far from the ground, even to the very top, waxing by degrees lesser and shorter still as they grow higher, the tree bearing the form or the shape of a pyramid or sharp pointed steeple; these compass the body round in manner of a circle, and are so orderly placed by degrees, as a man may climb up by them to the very top as by a ladder.

The leaves be small and round like those of the Pine tree, but shorter and not so sharp pointed; all the cones or clogs are far shorter and thicker than those of the Fir tree, compact of soft scales, not hard, which hang not downwards, but stand upright upon the boughs, whereunto also they are so strongly fastened as they can hardly be plucked off without breaking of some part of the branches, as Bellonius writes. The timber is extreme hard, and neither rots nor waxes old. There is no worms nor rottenness can take the hard matter or heart of this wood, which is very odoriferous and somewhat red. Solomon, King of the Jewes, did therefore build God's temple in Jerusalem of Cedar wood. The Gentiles were wont to make their Divels or images of this kind of wood, that they might last the longer.

The Place and Time.

The Cedar trees grow upon the snowy mountains, as in Syria upon mount Libanus, on which there remain some even to this day, says Bellonius, planted, as is thought, by Solomon himself. They are likewise found on the mountains Taurus and Amanus, in cold and stony places. The merchants of the factory at Tripolis told me that the Cedar grows upon the declining of the mountain Libanus, near unto the hermitage by the city Tripolis in Syria. They that dwell in Syria use to make boats thereof for want of the Pine tree. The Cedar tree remains always green, as other trees which bear such manner of fruit. The timber of the Cedar tree, and the images and the other works made thereof, seem to sweat and send forth moisture in moist and rainy weather, as do likewise all that have an oily juice, as Theophrastus writes.

The Names.

This huge and mighty tree is called in Latin, *cedrus.* In English, Cedar and Cedar tree. Pliny writes that the liquor of the Torch pine is named *cedrium.*

The Temperature and Vertues.

Cedar is of temperature hot and dry with such an exquisite tenuity and subtlety of parts, that it seems to be hot and dry in the fourth degree, especially the Pitch or Rosin thereof. There issues out of this tree a Rosin like unto that which issues out of the Fir tree, very sweet in smell, of a clammy or cleaving substance, the which if you chew in your teeth it will hardly be gotten forth again, it cleaves so fast. At the first it is liquid and white, but being dried in the sun it waxes hard. If it be boiled in the fire, an excellent pitch is made thereof, called Cedar pitch.

The Ægyptians were wont to coffin and embalm their dead in Cedar and with Cedar pitch, although they used also other means, as Herodotus records. Cedar infused in vinegar and put into the ears, kills the worms therein, and being mingled with the decoction of Hyssop, appeases the sounding, ringing and hissing of the ears. It is good to kill nits and lice and such like vermin. It is also good for Lepers. Being put up underneath, it kills all manner of worms and draws forth the birth, as Dioscorides writes.

Of the prickly Cedar, or Cedar Juniper.

The Kindes.

The prickly Cedar is like to Juniper, and is called the small or little Cedar, for difference from the great and tall Cedar which brings Cones; and of this there are two kinds as Theophrastus and Pliny do testify, that is to say, one of Lycia and another crimson.

The Description.

The Crimson or prickly Cedar seems to be very like to the Juniper tree in body and boughs, which are writhed, knotty, and parted into very many wings. The substance of the wood is red, and sweet of smell like that of the Cypresse; the tree is covered over with a rugged bark. The leaves be narrow and sharp pointed, harder than those of Juniper, sharper and more pricking, and standing thinner upon the branches. The fruit or berry is sometimes as big as a Hasell nut, or, as Theophrastus says, of the bigness of Myrtle berries; and being ripe it is of a reddish yellow, or crimson colour, sweet of smell, and so pleasant in taste as even the country-men now and then do eat of the same with bread.

The other low Cedar which grows in Lycia is not so high as the former, having likewise a writhed body as big as a man's arm, full of boughs. The bark is rough, yellow without and red within. The leaves

oxycedrus phœnicia.
Crimſon prickly Cedar.

oxycedrus lycia.
Rough Lycian Cedar.

stand thicker, like at the first to those of Juniper, but yet somewhat shorter, and in the third or fourth year thicker, long and round withal, coming near to the leaves of the Cypress tree, or of the second Savine, that is, blunt and not pricking at all, which being bruised between the fingers do yield a very pleasant smell, for does one and the self same plant bring forth below sharp and prickly leaves, and above thick and blunt ones, as that notable learned Herbarist Clusius has most diligently observed. The fruit or berry is round like that of Juniper, of colour yellow when it is ripe, inclining to a red, in taste somewhat bitter, but sweet of smell.

There is another Lycian Cedar that also has Cypresse-like leaves, not unlike those of the last described, yet somewhat thicker and broader. The fruit is also much larger, being as big as Hasell nuts, and of a red or scarlet colour, whence Lobel calls it *cedrus phœnicia altera.*

The Place and Time.

The prickly Cedar with the crimson colour comes up higher and greater in certain places of Italy, Spaine, and Asia, and other Countries; for that which grows on mount Garganus in Apulia is much higher and broader than those that grow elsewhere, and brings forth greater berries, of the bigness of an hasell nut, and sweeter, as that most diligent writer Bellonius reports. Carolus Clusius shows that the prickly Cedar and the Juniper tree be of so great a growth in diverse places of Spaine, as he has observed, as that the body of them is as thick as a man.

The Lycian Cedar is found in Provence of France, not far from Massilia, and grows in a great part of Greece, in Illyricum and Epirum.

Both of them are always green and in winter also full of fruit, by reason that they continually bring forth berries, as when the old do fall, new come in their places. In the spring grow up new buds and beginnings of berries. In autumn they wax ripe the second year, as do the berries of Juniper.

cedrus lycia altera.

The other Lycian Cedar.

The Names.

They are called in Latine, *minores,* and *humiles cedri,* little and low Cedars, for difference from the tall and great Cedar which bear cones.

The former is named in Latine, *oxycedrus* and *cedrus punica;* in English, Prickly Cedar and Crimson Cedar. The Spaniards call this also *enebro,* even by the same name which they give to Juniper, wherein they are thought to imitate diverse of the old Writers.

The other with the blunt leaf is named of Pliny, *lycia cedrus.* In Provence of France, *morveine.* Diverse name this *sabina,* and use it in stead of *Savine,* which they want.

The fruit of this Cedar is named by Theophrastus, *cedris;* notwithstanding *cedrus,* as he himself does also testify, is a certain little shrub which never grows to a tree.

The gum or liquor which issues forth of the prickly Cedar is also called *vernix,* and is sold in stead thereof.

The Temperature and Vertues.

The little Cedar, as Galen writes, is hot and dry in a manner in the third degree. The matter or substance thereof is sweet of smell, like that of Juniper, and is used for perfumes and odoriferous smells together with the leaves.

Of the Chaste Tree.

The Description.

Vitex, or the Chaste tree, grows after the manner of a bushie shrub or hedge tree, having many twiggie branches, very pliant and easy to be bent without breaking, like to the willow. The leaves are for the most part divided into five or seven sections or divisions, much like the leaves of Hemp, whereof each part is long and narrow, very like unto the willow leaf but smaller. The flowers do grow at the uppermost parts of the branches, of a light purple or blew colour, and very sweet smell. The fruit is small and round, like unto the graines or cornes of pepper.

Lobel mentions another variety hereof that differs from the former only in that it has broader leaves, and these also snipt about the edges.

The Place.

Vitex grows naturally in Italy and other hot regions, by water courses and running streams. I have it growing in my garden.

The Time.

Vitex begins to recover his last leaves in May, and the flowers come forth in August.

The Names.

In Latine, *salix marina* or *salix amerina,* and *piper agreste;* in High Dutch, *keuschbaum;* in Low Dutch and also of the Apothecaries, *agnus castus;* of the Italians, *vitice, agno casto;* in Spanish, *gattile casto;* in English, Chaste tree, Hempe tree, and of diverse, *agnus castus.*

The Temperature and Vertues.

The leaves and fruit of *agnus castus* are hot and dry in the third degree. They are of very thin parts, and waste or consume winde.

Agnus castus is a singular medicine and remedy for such as would willingly live chaste, for it withstands all uncleanness, or desire to the flesh, consuming and drying up the seed of generation, in what sort soever it be taken, whether in powder only, or the decoction drunk, or whether the leaves be carried about the body; for which cause it was called *castus,* that is to say, chaste, cleane, and pure.

The seed of *agnus castus* drunken, drives away and dissolves all windinesse of the stomacke, opens and cures the stoppings of the liver and spleen, and in the beginning of dropsies, it is good to be drunk in wine in the quantity of a dram.

The leaves stamped with butter, dissolve and assuage the swellings of the genitories and cods, being applied thereto.

The decoction of the herbe and seed is good against pain and inflammations about the matrix, if women be caused to sit and bathe their privy parts therein. The seed being drunk with Penny-royal

vitex latiore serrato folio.
Chaste tree with cut leaves.

vitex, siue agnus castus.
The Chaste tree.

brings down the menses, as it does also both in a fume and in a pessary. In a Pultis it cures the headache, the Phrenticke, and those that have the Lethargie are wont to be bathed herewith, oile and vinegar being added thereto.

The leaves used in a fume, and also strewed, drive away serpents; and being laid on do cure their bitings.

The seed laid on with water does heal the clifts or rifts of the fundament; with the leaves, it is a remedy for limbs out of joint and for wounds.

It is reported that if such as journey or travel do carry with them a branch or rod of *agnus castus* in their hand, it will keep them from Merry-galls and weariness.

Of the Cherrie tree.

cerasus vulgaris.

The common Englifh Cherrie tree.

The Description.

The English Cherrie tree grows to a high and great tree, the body whereof is of a mean bigness, which is parted above into very many boughs with a bark somewhat smooth, and of a brown crimson colour, tough and pliable. The substance or timber is also brown in the middle, and the outward part is somewhat white. The leaves be great, broad, long, set with veins or nerves, and slightly nicked about the edges. The flowers are white, of a mean bigness, consisting of five leaves, and having certain threads in the middle of the like colour. The cherries be round, hanging upon long stems or foot-stalks, with a stone in the midst which is covered with a pulp or soft meat; the kernel thereof is not unpleasant to the taste, though somewhat bitter.

The Time.

The Cherrie trees bloom in April; some bring forth their fruit sooner; some later. The red Cherries be always better than the black of their own kind.

The Names.

The Cherrie-tree is called in Latin, *cerasus;* in High Dutch, *kirschenbaum;* in Low Dutch, *kersenboome;* in French, *cerisier;* in English, Cherrie-tree.

The Temperature and Vertues.

The best and principal Cherries be those that are somewhat sour. Those little sweet ones which be wild and soonest ripe be the worst: they contain bad juice, they very soon putrefy, and do ingender ill blood, by reason whereof they do not only breed worms in the belly, but troublesome agues, and often pestilent fevers, and therefore in well-governed common wealths, it is carefully provided that they should not be sold in the markets in the plague time.

Generally, all the kinds of Cherries are cold and moist of temperature, although some more cold and moist than others; the which being eaten before meat do soften the belly very gently. They are unwholesome either to moist and rheumatic bodies, or for unhealthy and cold stomachs.

The distilled water of Cherries is good for those that are troubled with heat and inflammations in their stomachs, and prevails against the falling sickness given mixed with wine. Many excellent tarts and other pleasant meats are made with Cherries, sugar, and other delicate spices, whereof to write were to small purpose.

The gum of the Cherrie tree taken with wine and water, is reported to help the stone; it may do good by making the passages slippery, and by tempering and allaying the sharpness of the humours, and in this manner it is a remedy also for an old cough. Dioscorides adds that it makes one well coloured, clears the sight, and causes a good appetite to meat.

Of the Chestnut Tree.

castanea.
Cheſtnut tree.

The Description.

The Chestnut tree is a very great and high tree. It casts forth very many boughs. The body is thick and sometimes of so great compass as that two men can hardly fathom it. The timber or substance of the wood is sound and durable. The leaves be great, rough, wrinkled, nicked in the edges, and greater than the particular leaves of the Walnut tree. The blowings or catkins be slender, long, and green. The fruit is enclosed in a round, rough and prickly husk, like to a hedge-hog or urchin, which opening it self does let fall the ripe fruit or nut. This nut is not round, but flat on one side, smooth, and sharp-pointed. It is covered with a hard shell, which is tough and very smooth, of a dark brown colour. The meat or inner substance of the nut is hard and white, and covered with a thin skin which is under the shell.

The Horse Chestnut grows likewise to be a very great tree, spreading his great and large arms or branches far abroad, by which means it makes a very good cool shadow. These branches are garnished with many beautiful leaves, cut or divided into five, six, or seven sections or divisions, like to the Cinkfoile, or rather like the leaves of *ricinus*, but bigger. The flowers grow at the tops of the stalks, consisting of four small leaves like the Cherry blossom, which turn into round, rough, prickly heads like the former, but more sharp and harder. The nuts are also rounder. The fruit is contained in a prickly husk that opens in three parts, and it is rounder and not so sharp-pointed as the ordinary Chestnut, neither under the outer coat has it any peeling within as the other has, neither is it of so good a taste.

The Place.

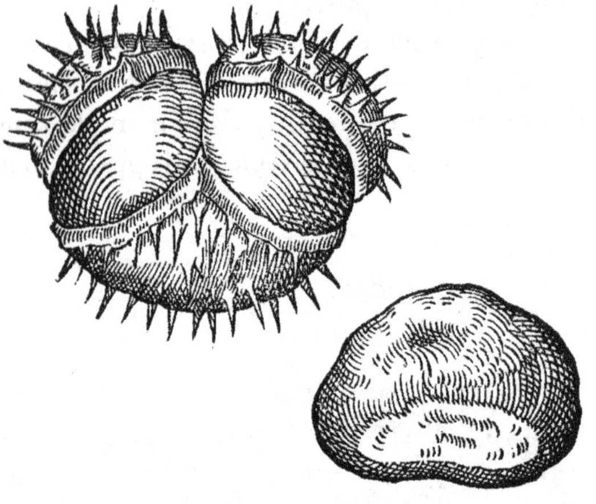

The first grows on mountains and in shadowy places, and many times in the valleys. They love a soft and black soil. There be sundry woods of Chestnuts in England, as a mile and a half from Feversham in Kent, and in sundry other places. In some countries they be greater and pleasanter; in others, smaller and of worse taste. The Horse Chestnut grows in Italy and in sundry places of the East countries. It is now growing with Mr Tradescant in South Lambeth.

The Time.

The blowings or aglets come forth with the leaves in April; the nuts later, and be not ripe till Autumn.

The Names.

The Chestnut tree bears the name of the nut, both in Greek and Latin. In High Dutch, *kastanibaum;* in French, *castaignier;* in English, Chestnut tree. The nut is called in Latin, *castanea;* in High Dutch, *kesten;* in Italian, *castagne;* in French, *chastaigne;* in Spanish, *morones;* in English, Chestnut. The greater nuts be named of the Italians, *marroni;* of the Frenchmen and of diverse base Almaines, *marons.*

The Horse Chestnut is called in Latin, *equina castanca;* in English, Horse Chestnut, for that the people of the East countries do with the fruit thereof, cure their horses of the cough, shortness of breath, and such like diseases.

The Temperature and Vertues.

Our common Chestnuts are very dry and binding, and be neither hot nor cold, but in a mean between both; yet have they in them a certain windiness, and by reason of this, unless the shell be first cut, they skip suddenly with a crack out of the fire whilst they be roasting.

Of all the Acorns, says Galen, the Chestnuts are the chiefest, and do only of all the wild fruits yield to the body commendable nourishment; but they slowly descend, they be hardly concocted, they make a thick blood, and ingender wind. They also stay the belly, especially if they be eaten raw.

Some affirm that of raw Chestnuts dried, and afterwards turned into meal, there is made a kind of bread. Yet it must needs be that this should be dry and brittle, hardly concocted, and very slow in passing through the belly; but this bread may be good against the laske and bloody flix.

An Electuarie of the meal of Chestnuts and honey is very good against the cough and spitting of blood. The bark of the Chestnut tree boiled in wine and drunk, stops the laske, the bloody flix, and all other issues of blood.

Of the Cinnamon, or Canell tree.

The Description.

canellæ folium, bacillus, & cortex.
The leafe, barke, and trunk of the Cinnamon tree.

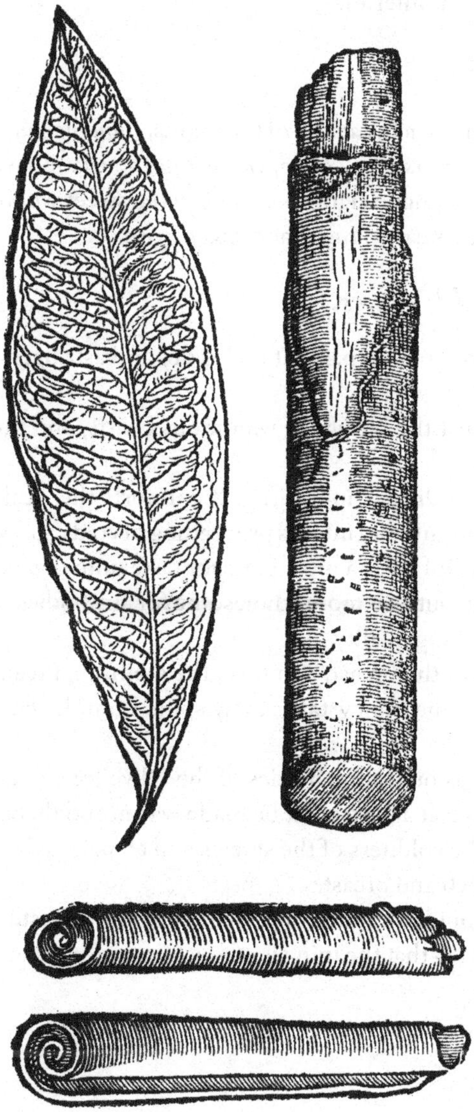

The tree which has the Cinnamon for his bark is of the stature of an Olive tree, having a body as thick as a man's thigh, from which the Cinnamon is taken; but that taken from the smaller branches is much better; which branches or boughs are many, and very straight; whereon do grow beautiful leaves in shape like those of the Orenge tree, and of the colour of the Bay leaf. Among these pleasant leaves and branches come forth many fair white flowers, which turn into round black fruit or berries, of the bigness of a Hasell nut, or the Olive berry, out of which is pressed an oile that has no smell at all until it be rubbed and chafed between the hands. The trunk or body, with the greater arms or boughs of the tree, are covered with a double or two-fold bark, like that of *suber,* the Cork tree. The innermost whereof is the true and pleasant Cinnamon, which is taken from the tree and cast upon the ground in the heat of the Sun; through the heat thereof it turns and folds it self round together, as we daily see by viewing the thing it self. This tree being thus peeled, recovers a new bark in the space of three years, and is then ready to be disbarked as afore. That Cinnamon which is of a pale colour, has not been well dried in the Sun. That of a fair brown colour is best; and that which is blackish, has been too much dried and also has taken some wet in the time of drying.

The Place.

The chiefest places where the trees do grow that bear Cinnamon, are Zeilan and Malavar; but those of Zeilan are the best. They grow in other of the Molucca islands, as Java, the greater and the less, and also in Mindanoa, for the most part upon mountains.

The Time.

The Cinnamon tree grows green Winter and Summer, as do all the other trees of the Moluccas and East Indies, for the most part; the boughs whereof are cut off at seasonable times by the express commandment of the King of the country; and not before he has appointed the time.

There has been some controversy among writers concerning the tree whose bark is *cassia,* and that tree that bears Cinnamon, making them both one tree; but that opinion is not to be received. For there is a great difference between them, as there is betwixt an Oke and a Chestnut tree; for the tree whose bark is *cassia* is doubtless a bastard kind of Canell or Cinnamon. In show it is very like, but in sweetness of smell and other circumstances belonging to Cinnamon, far inferior.

The Names.

Cinnamon is called in Italian, *canella;* in Spanish, *canola;* in French, *canelle;* of the Latins, *cinnamomum;* of the Arabians, *darseni* and as some say, *quersaa;* others, *querse.* In Zeilan, *cuurde;* in the island Java they name it *cameaa;* in Malavar, *cais mains,* which in their tongue signifies *dulce lignum,* sweet wood. In English, Cinnamome, Cinnamon, and Canell. The other is called *cassia* and *cassia lignea.*

The Temperature and Vertues.

Dioscorides writes that Cinnamon has power to warm and is of thin parts. It is also dry and astringent, it provokes urine, clears the eyes, and makes sweet breath.

The decoction brings down the menses, prevails against the bitings of venomous beasts, and the inflammation of the intestines and reines.

The distilled water hereof is profitable to many, and for diverse infirmities. It comforts the weak, cold, and feeble stomach; eases the pains and frettings of the guts and intrails proceeding of cold causes; it amends the evill colour of the face, makes sweet breath, and gives a most pleasant taste unto diverse sorts of meats, and makes the same not only more pleasant, but also more wholesome for any bodies of what constitution soever they be, notwithstanding the binding quality.

The oile, drawn chimically, prevails against the pains of the breast, comforts the stomach, breaks windiness, causes good digestion, and being mixed with some honey, takes away spots from the face, being anointed therewith.

Out of the berries of this tree is drawn by expression, as out of the berries of the Olive tree, a certain oile, or rather a kind of fat like butter, without any smell at all, except it be made warm, and then it smells as the Cinnamon does, and is much used against the coldness of the sinews and all pains of the joints, and also the pains and distemperature of the stomach and breast.

To write as the worthiness of the subject requires, would ask more time than we have to bestow upon any one plant; therefore these few shall suffice, knowing that the thing is of great use among many and known to most.

malus limonia.

The Limon tree.

Of the Citron, Limon, and Orange trees.

The Description.

The Citron tree is not very great, having many boughs or branches, tough and pliable, covered with a green bark; whereon do grow green leaves, long, somewhat broad, very smooth, and sweet of smell like those of the Bay tree; among which come forth here and there certain prickles, set far in sunder; from the bosom whereof come forth small flowers consisting of five little leaves, of a white colour tending to purple, with certain threads like hairs growing in the middle. The fruit is long, greater many times than the Cucumber, often lesser, and not much greater than the Limon. The bark or rind is of a light golden colour, set with diverse knobs or bumps, and of a very pleasant smell. The pulp or substance next unto it is thick, white, hard, having a kind of aromatical or spicy smell, almost without any taste at all. The softer pulp within that, is not so firm or solid but more spongy, and full of a sour juice; in which the seed lies hid, greater and thicker than a grain of Barley.

The Limon tree is like unto the Citron tree in growth, thorny branches, and leaves of a pleasant sweet smell, like those of the Bay tree. The flowers hereof are whiter than those of the Citron tree, and of a most sweet smell. The fruit is long and thick, lesser than the Citron; the rind is yellow, somewhat bitter in taste and sweet of smell. The pulp is white, more in quantity than that of the Citron, respecting the bigness, in the middle part whereof is contained more soft spongy pulp, and fuller of sour juice. The seeds are like those of the Citron.

The Orange tree grows up to the height of a small Pear tree, having many thorny boughs or branches like those of the Citron tree. The leaves are also like those of the Bay tree & they are of a sweet smell; the flowers are white, of a most pleasant sweet smell also. The fruit is round like a ball, every circumstance belonging to the form is very well known to all. The taste is sour, sometimes sweet, and often of a taste between both. The seeds are like those of the Limon.

The Place and Time.

The Citron, Limon, and Orange trees do grow especially on the sea coasts of Italy, and on the islands of the Adriatic, Tyrrhenian, and also Ægean seas, and likewise on the main land near unto meres and great lakes. There is also great store of them in Spaine, but in places especially joining to the sea, or not far off. They are also found in certain provinces of France which lie upon the Midland sea.

These trees be always green and do, as Pliny says, bear fruit at all times of the year, some falling off, others waxing ripe, and others newly coming forth.

The Names.

The Citron is named in Latin, *malum medicum* and *malum citrium* and *citromalum.* The Apothecaries call these apples, *citrones.* In High Dutch, *citrin opffell;* in Low Dutch, *citroenen;* in Italian, *citroni;* in Spanish, *cidras;* in French, *citrons;* in English, Citron apple and Citron.

The Limon is called in Latin, *limonium malum;* in shops, *limones;* in French, *limons;* in Low Dutch, *limonen;* in English, Limon and Lemon.

Oranges are named in Latin, *malum anarantium,* and of some, *aurantium,* of others, *aurengium,* of the yellow colour of gold. In High Dutch, *pomeranken;* in Low Dutch, *araengie appelen;* in French, *pommes d'orenges;* in Spanish, *naransas;* in English, Oranges.

The Temperature and Vertues.

All these fruits consist of unlike parts, and much differing in faculty. The rinds are sweet of smell, bitter, hot, and dry. The white pulp is cold, and contains within it a gross juice, especially the Citron. The inner substance or pap is sour, as of the Citrons and Limons, cold and dry, with thinness of parts. The seed, because it is bitter, is hot and dry. The rind of the Citron is good against all poisons, for which cause it is put into treacles and such like confections. It is good to be eaten against a stinking breath. The white pulp is now and then eaten, but very hardly concocted, and ingenders a gross, cold, and phlegmaticke juice; but being condite with sugar, it is both pleasant to taste and easy to be digested, more nourishing, and less apt to binding or stopping. Galen reports that the inner juice of the Citron was not wont to be eaten, but it is now used for sauce, and being so used, it represses choler which is in the stomach, and procures appetite. It is excellent good also to be given in vehement and burning fevers, and against all pestilent and venomous or infectious diseases; it comforts the heart, cools the inward parts, cuts, divides, and makes thin, gross, tough and slimy humours.

Of this foresaid sharp juice there is a syrup prepared, which is called in shops, *syrupus de acetositate citri,* very good against the foresaid infirmities. Such a syrup is also prepared of the sharp juice of Limons, of the same quality and operation, so that in stead of the one, the other will serve very well.

Two ounces of the juice of Limons, mixed with the like quantity of the spirit of wine, and drunk at the first approach of the fit of an ague, takes away the shaking presently. The medicine seldom fails at the second time of the taking thereof, perfectly to cure the same; but never at the third time, provided that the patient be covered warm in a bed, and caused to sweat.

The seed of all these does kill worms in the belly and drives them forth. It does also mightily resist poison, and is good for the stinging of scorpions if it be inwardly taken.

Of the Clove tree.

caryophylli very clusij.

The true forme of the Clove tree.

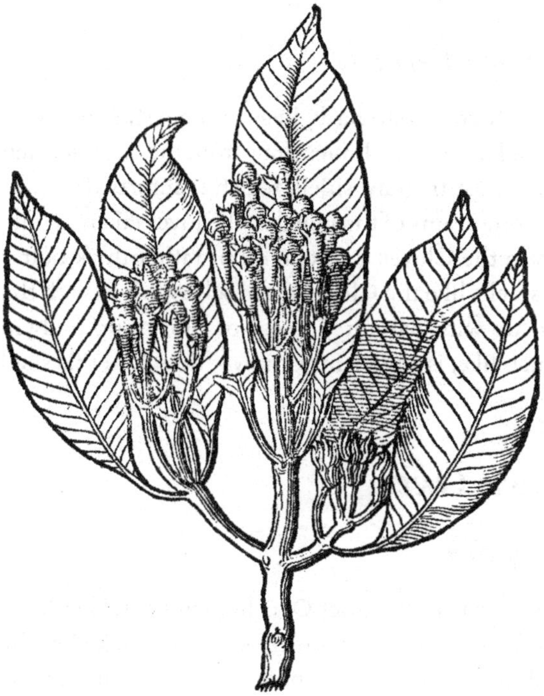

The Description.

The Clove tree grows great in form like unto the Bay tree, the trunk or body whereof is covered with a russet bark. The branches are many, long, and very brittle, whereupon do grow leaves like those of the Bay tree, but somewhat narrower; amongst which come the flowers, white at the first, after of a greenish colour, waxing of a dark red colour in the end; which flowers are the very Cloves when they grow hard. After, when they be dried in the Sun, they become of that dusky black colour which we daily see, wherein they continue. For those that we have in estimation are beaten down to the ground before they be ripe, and are suffered there to lie upon the ground until they be dried thoroughly, where there is neither grass, weeds, nor any other herbs growing to hinder the same, by reason the tree draws unto it self for his nourishment all the moisture of the earth a great circuit round about, so that nothing can there grow for want of moisture, and therefore the more convenient for the drying of the cloves.

Contrariwise, that gross kind of cloves which has been supposed to be the male, are nothing else than fruit of the same tree tarrying there until it fall down of it self unto the ground, where by reason of his long lying, and meeting with some rain in the mean season, it loses the quick taste that the others have. Some affirm that the flowers hereof surpass all other flowers in sweetness when they are green; and hold the opinion that the hardened flowers are not the Cloves themselves (as we have written), but think them rather to be the seat or husk wherein the flowers do grow. The greater number hold the former opinion; and further, that the trees are increased without labour, graffing, planting, or other industry, but by the falling of the fruit; which bear fruit within eight years after they be risen up, and so continue bearing for an hundred years together, as the inhabitants of that country do affirm.

The Place and Time.

The Clove tree grows in some few places of the Molucca Islands, as in Zeilan, Java the greater and the lesser, and in diverse other places. The cloves are gathered from the fifteenth of September unto the end of February, not with hands, as we gather Apples, Cherries and such like fruit, but by beating the tree, as Walnuts are gotten, as we have written in the description.

The Names.

It is called in Latin, *caryophyllus* and *clavus;* in French, *clou de gyrofle;* in High Dutch, *naegel;* in Spanish, *clavo de especia;* of the Indians, *calafur;* in the Moluccas, *changue;* in English, Clove tree and Cloves.

The Temperature and Vertues.

Cloves are hot and dry in the third degree. They strengthen the stomach, liver, and heart; help digestion; and provoke urine. The Portugal women that dwell in the East Indies draw from the Cloves when they be yet green, a certain liquor by distillation, of a most fragrant smell, which comforts the heart and is of all cordials the most effectual. The weight of four drams of the powder of Cloves taken in milk procures the act of generation. There is extracted from the Cloves a certain oil or rather thick butter of a yellow colour; which being chafed in the hands smells like the Cloves them selves, wherewith the Indians do cure their wounds and other hurts, as we do with Balsam. The use of Cloves, not only in meat and medicine, but also in sweet powder and such like, is sufficiently known, therefore this shall suffice.

Of the Corke Oke.

The Description.

The Corke tree is of a middle bigness like unto *ilex,* or the barren scarlet Oke, but with a thicker body and fewer boughs. The leaves be for the most part greater, broader, rounder, and more nicked in the edges. The bark of the tree is thick, very rugged, and full of chinks or crannies that cleave and divide it self into pieces, which unless they be taken away in due time do give place to another bark growing underneath, which when the old is removed is marvellous red, as though it were painted with some colour. The Acorn stands in a cup, which is great, bristled, rough, and full of prickles.

The Corke tree with narrow leaves grows likewise to the height and bigness of a great tree, the trunk or body whereof is covered with a rough and scabbed bark of an overworn blackish colour, which cleaves and casts his coat when the inner bark grows somewhat thick. The branches are long, tough and flexible, easy to be bowed any way; whereupon do grow narrow leaves like those of the precedent, but longer, and little or nothing indented about the edges. The fruit grows in small cups as the Acorns do; they are lesser than those of the other kind, as is the rest of the tree, which is the chiefest difference.

The Place and Time.

It grows in the country of Aquitania, near to the mountains called Pyrenæi. It also grows plentifully in the kingdom of Spaine, differing somewhat from that of Aquitania, as Clusius declares. It is likewise found in Italy, and that in the territory of Pisa, with a longer leaf and sharper-pointed; and about Rome with a broader leaf and cut in the edges like a saw, and rougher, as Matthiolus testifies. The leaves are always green in Spaine and Italy; about the Pyrenæan mountains they fall away in Winter.

The Names.

The tree is called in Latin, *suber;* in French, *liege;* in Italian, *sugaro;* the same names do also belong to the bark. The Spaniards call the tree *alcornoque;* the English men, Cork tree, whereupon the Low-country men and English men call the bark, Cork.

suber latifolium & angustifolium.
The Corke tree with broad and narrow leaves..

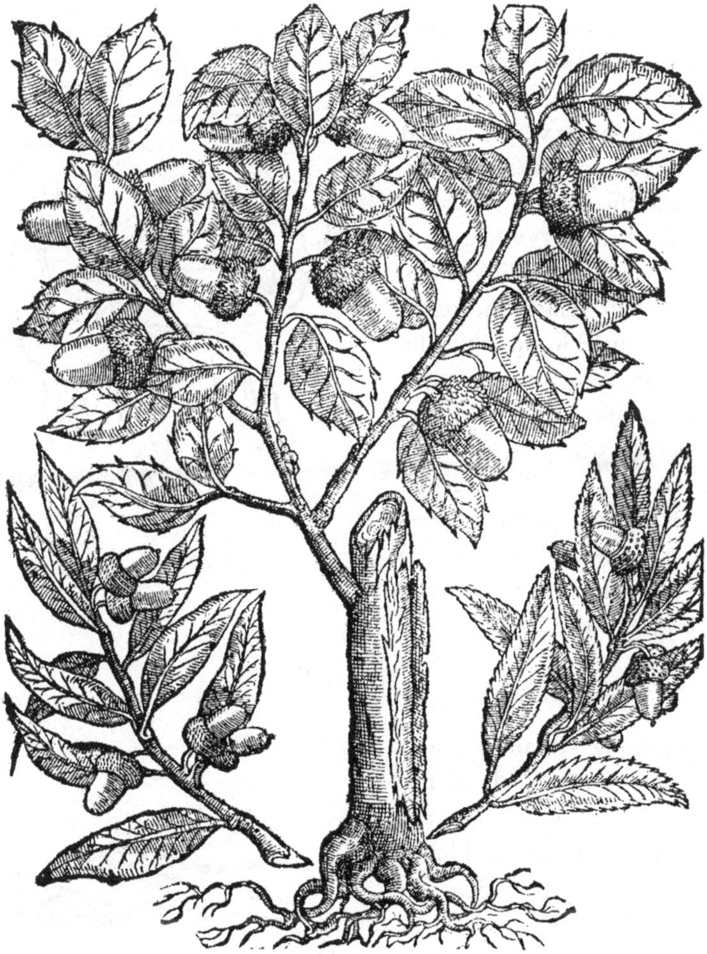

The Temperature and Vertues.

This bark does manifestly dry, with a binding faculty. Being beaten to powder and taken in water, it stanches bleeding in any part of the body. The Cork which is taken out of wine vessels, being burnt, makes ashes which do mightily dry, and are mixed in compositions devised against the bloody flix. Cork is also profitable for many things: it is used about the anchors of ships, fisher's nets, and to stop vessels with, and in Winter for women's shoes, which use remains with us even to this day. Fishermen hang this bark upon the wings of their nets for fear of sinking; and shoe-makers put it in shoes and pantosles for warmness sake.

Of the Cornell tree.

The Description.

The tame Cornell tree grows sometime of the height and bigness of a small tree, with a great number of springs. It is covered with a rugged bark. The wood or timber is very hard and dry, without any great quantity of sap therein. The leaves are like unto the Dog berry leaves, crumpled, rugged, and of an overworn colour. The flowers grow in small bunches before any leaves do appear, of colour yellow and of no great value (they are so small) in shew like the flowers of the Olive tree. Which being faded, there come forth small long berries which at the first be green, and red when they be ripe, of an austere and harsh taste with a certain sourness. Within this berry is a small stone, exceeding hard, white within like that of the Olive, whereunto it is like both in the fashion and oftentimes in the bigness of the fruit.

The Place and Time.

This grows in most places of Germanie without manuring. It grows not wild in England, but there be sundry trees of them growing in the gardens of such as love rare and dainty plants, whereof I have a tree or two in my garden.

cornus mas.

The male Cornel tree.

The tame Cornell tree flowers sometime in February, & commonly in March, and afterwards the leaves come forth as an untimely birth. The berries or fruit are ripe in August.

The Names.

The Latines call it *cornus.* In High Dutch, *cornelbaum;* in Low Dutch, *cornoele boom;* the Italians, *corniolo;* in French, *cornellier;* in Spanish, *cornizolos;* in English, the Cornell tree and the Cornelia tree; of some, long Cherrie tree.

The fruit is named in Latine, *cornum;* in High Dutch, *cornell;* in Low Dutch, *cornoele;* in Italian, *cornole;* in English, Cornel berries and Cornelian Cherries.

This is *cornus mas Theophrasti,* or Theophrastus his male Cornell tree, for he sets down two sorts of the Cornell trees, the male and the female. He makes the wood of the male to be found as in this Cornell tree; which we both for this cause and for others also have made to be the male. The female is that which is commonly called *virga sanguinea,* or Dog's berry tree, and *cornus sylvestris,* or the wilde Cornell tree, of which we will treat in the next Chapter following.

The Temperature and Vertues.

The fruit of the Cornell tree has a very harsh or choking taste. It cools, dries, and binds, yet may it also be eaten, as it is oftentimes.

It is a remedie against the laske and bloudy flix, it is hurtful to a cold stomacke, and increases the rawness thereof. The leaves and tender crops of the tree are likewise of a harsh and choking taste, and do mightily dry.

They heal green wounds that are great and deep, especially in hard bodies, but they are not so good for small wounds and tender bodies, as Galen writes.

Of the female Cornell or Dog-Berry tree.

The Description.

That which the Italians call *virga sanguinea,* or the bloudy Rod, is like to the Cornell tree, yet it grows not into a tree but remains a shrub. The young branches thereof are jointed, and be of an obscure red purple. They have within a white spongie pith like that of Elder, but the old stalks are hard and stiff, the substance of the which is also white, and answerable to those of the Cornell tree. The leaves are also like, the middle rib whereof as also the brittle foot-stalks are somewhat reddish; at the top whereof stand white flowers in spoky rundles, which turn into berries, green at the first, and of a shining black colour when they be ripe, in taste unpleasant, and not cared for of the birds.

The Place.

This shrub grows in hedges and bushes in every country of England.

cornus foemina.
The Dog-berry tree.

The Time.

The flowers come forth in the Spring in the month of April. The berries are ripe in Autumn.

The Names.

The Italians do commonly call it *sanguino* and *sanguinello.* Petrus Crescentius terms it *sanguinus* and Matthiolus, *virga sanguinea.* Pliny has written a little of *virga sanguinea;* neither is *virga sanguinea,* says he, counted more happy, the inner bark whereof does break open the scarres which they before have healed. It is a hard thing, or peradventure a rash part, to affirm these few words: that Pliny his *virga sanguinea* is the same that the Italian *sanguino* is.

This is called in High Dutch, *hartriegel;* in Low Dutch, *wilde cornoelle,* that is to say, *cornus sylvestris* or wilde Cornell tree; and in French, *cornellier sauvage.* In English, Hound's tree, Hound's berry, Dog's berry tree, Pricke-Timber. In the North country they call it Gaten tree or Gater tree; the berries whereof seem to be those which Chaucer calls Gater berries.

The Temperature.

The berries hereof are of unlike parts; for they have some hot, bitter, and cleansing, and very many cold, dry, harsh, and binding; yet they have no use in medicine.

The Vertues.

Matthiolus writes that out of the berries first boiled, and afterwards pressed, there issues an oyle which the Anagnian country people do use in lamps; but it is not certain, not very like, that the bark of this wilde Cornell tree has that operation which Pliny reports of *virga sanguinea,* for he says, as we have already set down, that the inner bark thereof does break and lay open the scars which they before have healed.

Of the Cypresse tree.

The Description.

The tame or manured Cypresse tree has a long thick and straight body, whereupon many slender branches do grow, which do not spread abroad like the branches of other trees, but grow up alongst the body, yet not touching the top. They grow after the fashion of a steeple, broad below and narrow toward the top. The substance of the wood is hard, sound, well compact, sweet of smell, and somewhat yellow, almost like the yellow Saunders, but not altogether so yellow; neither does it rot nor wax old, nor cleaves or chops it self.

cupressus sativa & sylvestris.
The Garden & wild Cypreſſe tree.

The leaves are long, round like those of Tamariske, but fuller of substance. The fruit or nuts do hang upon the boughs, being in manner like to those of the Larch tree, but yet thicker and more closely compact. Which being ripe do of themselves part in sunder and then falls the seed, which is shaken out with the wind. The same is small, flat, very thin, of a swart ill favoured colour, which is pleasant to Ants or Pismires, and serves them for food.

Of this diverse make two kinds, the female and the male; the female barren and the male fruitful. Theophrastus reports that diverse affirm the male to come of the female. The Cypresse yields forth a certain liquid Rosin, like in substance to that of the Larch tree, but in taste marvellous sharp and biting.

The wilde Cypresse, as Theophrastus writes, is a high tree, and always green, so like to the other Cypresse as it seems to be the same both in boughs, body, leaves, and fruit, rather than a certain wilde Cypresse. The matter or substance of the wood is sound, of a sweet smell, like that of the Cedar tree, which rots not. There is nothing so crisped as the root, and therefore they use to make precious and costly works thereof.

[Thomas Johnson writes: I know no difference between the wilde and tame Cypresse of our Author, but in the handsomnesse of their growth, which is helped somewhat by art.]

The Place.

The tame and manured Cypresse grows in hot countries, as in Candy, Lycia, Rhodes, and also in the territorie of Cyrene. It is reported to be likewise found on the hills belonging to Mount Ida, and on the hills called Leuci, that is to say white, the tops whereof be always covered with snow. Bellonius denies it to be found upon the tops of the hills, but in the bottoms on the rough parts and ridges of the hills. It grows likewise in diverse places of England where it has been planted, as at Sion a place near London, sometime a house of Nuns. It grows also at Greenwich, and at other places, and likewise at Hampsted in the garden of Mr Wade, one of the Clerks of her Majesties privy Council.

The wilde kind of Cypresse tree grows hard by Ammon's Temple, and in other parts of the country of Cyrene upon the tops of mountains, and in extreme cold countries. Bellonius affirms that there is found a certain wilde Cypresse also in Candy, which is not so high as other Cypresse trees, nor grows sharp toward the top, but is lower and has his boughs spread flat, round about in compass. He says the body thereof is also thick, but whether this be that of which Theophrastus and Pliny make mention, we leave to consideration.

The Time.

The tame Cypres tree is always green; the fruit may be gathered thrice a year, in Januarie, May and September, and therefore it is surnamed *trifera*.

The wilde Cypresse tree is late, and very long before it buds.

The Names.

The tame Cypres is called in Latine, *cupressus;* in shops, *cypressus;* in Italian, *cypresso;* in French and Spanish, *cipres;* in High Dutch, *cipressenbaum;* in Low Dutch, *cypresse boom;* in English, Cypres and Cypres tree.

The fruit is called in Latine, *pilulæ cupressi, nuces cupressi,* and *galbuli;* in shops, *nuces cypressi;* in English, Cypres nuts or clogs. This tree in times past was dedicated to Pluto, and was said to be deadly, whereupon it is thought that the shadow thereof is unfortunate.

Theophrastus attributes great honor to this tree, showing that the roofs of old Temples became famous by reason of that wood, and that the timber thereof, of which the rafters are made, is everlasting, and it is not hurt there by rotting, cobweb, or any other infirmitie or corruption.

The Temperature.

The fruit and leaves of the Cypres are dry in the third degree, and astringent.

The Vertues.

The Cypres nuts being stamped and drunken in wine, as Dioscorides writes, stops the laske and the bloody flix, it is good against the spitting of blood and all other issues of blood.

They glue and heal up great ulcers in hard bodies; they safely and without harm soak up and consume the hid and secret moisture lying deep and in the bottom of weak and moist infirmities.

The leaves and nuts are good to cure the rupture, to take away the polypus, being an excrescence growing in the nose.

Some do use the same against carbuncles and eating sores, mincing them with parched Barley meale.

The leaves of Cypres boiled in sweet wine or Mead, helps the strangurie and difficulty of making water.

It is reported, that the smoke of the leaves does drive away gnats, and that the clogs do likewise.

The shavings of the wood laid among garments, preserves them from the moths. The rosin kills moths, little worms, and maggots.

Of the Date tree.

The Description.

The Date tree grows very great and high. The body or trunk thereof is thick, and covered with a scaly rugged bark, caused by the falling away of the leaves. The boughs grow only on the top, consisting of leaves set upon a woody middle rib like those of Reeds or Flags; the inner part of which rib or stalk is soft, light, hollow and spongy. Among the leaves come forth the flowers included in a long skinny membrane, as it were a sheath or hose, like that which covers the Floure-de-Luce before it be blown, which being opened of it self, white flowers start forth, standing upon short and slender foot-stalks,

palma.
The Date tree.

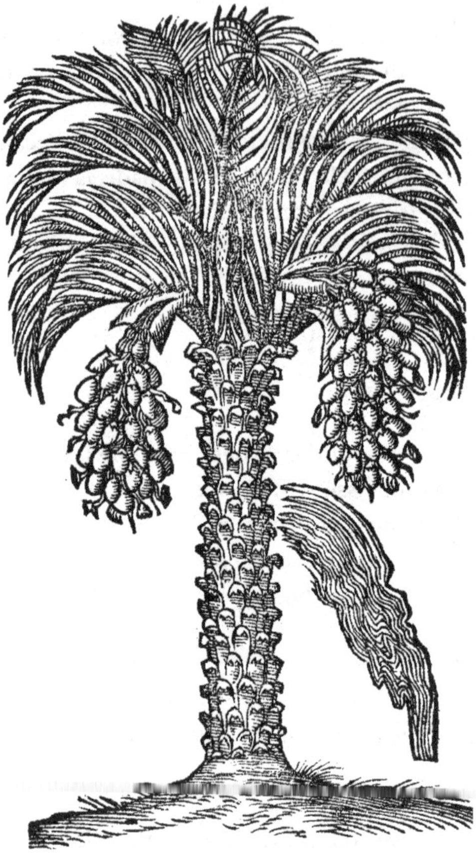

palmarum fructus & flores cum elate.
The fruit and flowers of the Date tree.

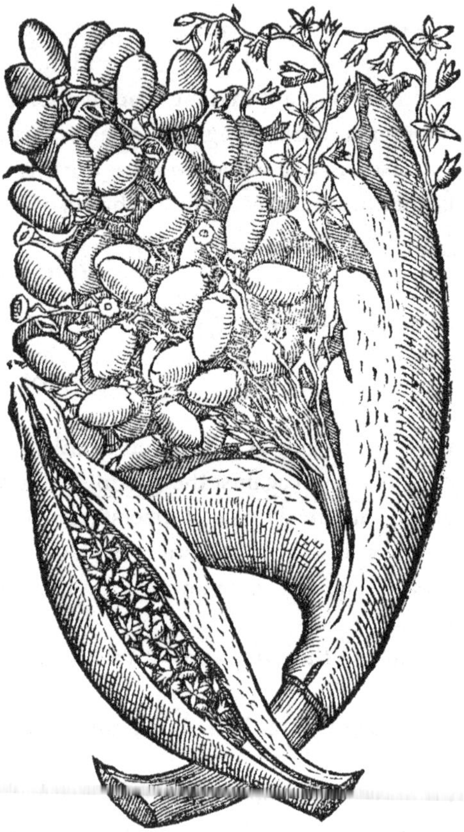

which are fastened with certain small filaments or thready strings like unto little branches; after which spring out from the same branches the fruit or Dates, which be in fashion long and round, in taste sweet, and many times somewhat harsh, of a yellowish red colour; wherein is contained a long hard stone which is in stead of kernel and seed; the which I have planted many times in my garden, and have grown to the height of three foot. But the first frost has nipped them in such sort that soon after they perished, notwithstanding my industry by covering them, or what else I could do for their succour.

The Place and Time.

The Date trees grow plentifully in Africa and Egypt, but those which are in Palestina and Syria be the best. They grow likewise in most places of the East and West Indies, where there be diverse sorts, as well wild as tame or manured. The Date tree is always green, and flowers in the Spring time. The fruit is ripe in September, and being then gathered they are dried in the Sun, that they may be the better both transported into other countries far distant, as also preserved from rotting at home.

The Names.

The tree is called in Latin, *palma;* in English, Date tree. The fruit is named in Latin, *glans palmarum,* or the fruit of the Date tree; and by one word, *palmula.* In shops, *dactyliss;* in High Dutch, *dattelen;* in Low Dutch, *dadelen;* in Italian, *dattoli;* in French, *dattes;* in Spanish, *tamaras* and *dattiles;* in English, Date.

The Temperature and Vertues.

All manner of Dates whatsoever are hard of digestion, and cause head-ache. The worser sort be those that be dry and binding, as the Egyptian dates; but the soft, moist and sweet ones are less hurtful. The Dates which grow in colder regions, when they cannot come to perfect ripeness, if they be eaten too plentifully, do fill the body full of raw humours, ingender wind, and oft times cause the leprosy. The best Dates, called in Latin, *caryotæ,* are good for the roughness of the throat and lungs.

There is made hereof both by the cunning confectioners and cooks, diverse excellent cordial, comfortable and nourishing medicines, and that procure lust of the body very mightily. The ashes of the Date stones have a binding quality and emplastick faculty; they heal pushes in the eyes, *staphylomata,* and falling away of the hair of the eye lids, being applied together with Spikenard. With wine it keeps proud flesh from growing in wounds. The leaves and branches of the Date tree do heal green wounds and ulcers, and refresh and cool hot inflammations. Galen, in his book of Medicines, mentions a composition called *diapalma* which is to be stirred with the bough of a Date tree in stead of a spature or a thing to stir with, for no other cause than that it may receive thereby some kind of astriction or binding force.

Of the Dragon tree.

The Description.

This strange and admirable tree grows very great, resembling the Pine tree, by reason it does always flourish, and has his boughs or branches of equal length and bigness, which are bare and naked, of eight or nine cubits long, and of the bigness of a man's arm; from the ends of which do shoot out leaves of a cubit and a half long, and full two inches broad, somewhat thick, and raised up in the middle, then thinner and thinner like a two-edged sword; among which come forth little mossy flowers which turn into berries of the bigness of Cherries, of a yellowish colour, round, light and bitter, covered with a threefold skin or film, wherein is to be seen the form of a Dragon, having a long neck or gaping mouth, the ridge or back armed with sharp prickles like the Porcupine. It has also a long tail, & four feet, very easy to be discerned; the figure of it we have set forth unto you according to the greatness thereof, because our words and meaning may be the better understood, and also the leaf of the tree in his full bigness, because it is impossible to be expressed in the figure.

The trunk or body of the tree is covered with a rough bark, very thin, and easy to be opened or wounded with any small tool or instrument; which being so wounded in the Dog days, bruised or bored, does yield forth drops of a thick red liquor, which of the name of the tree are called Dragon's tears, or *sanguis draconis,* Dragon's blood.

The Place and Time.

This tree grows in an island which the Portugals call Madera, and in one of the Canarie Islands called *Insula Portus Sancti,* and as it seems it was first brought out of Africke, although some are of a contrary opinion and say that it was first brought from Carthagena in America, by the Bishop of the same Province. The time of his growing we have touched in the description, where we said that it flourishes and grows green all the year.

draco arbor.
The Dragon tree.

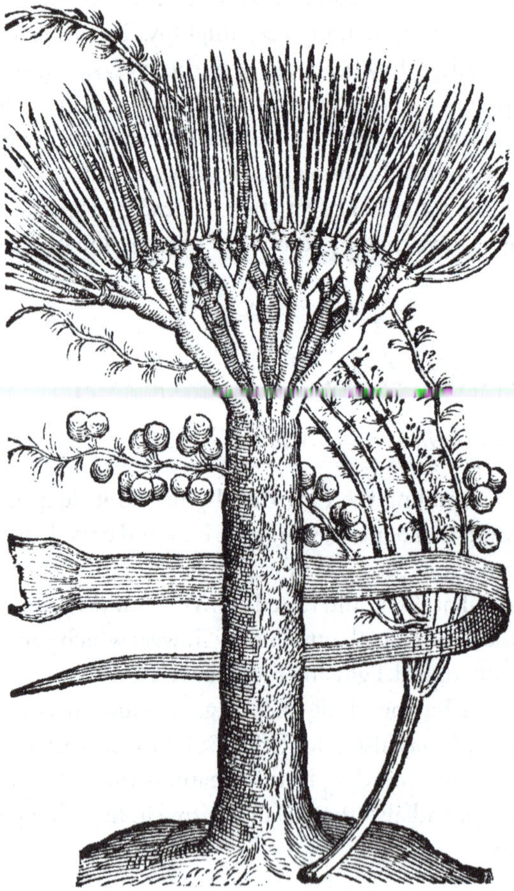

draconis fructus.
The Dragon tree fruit.

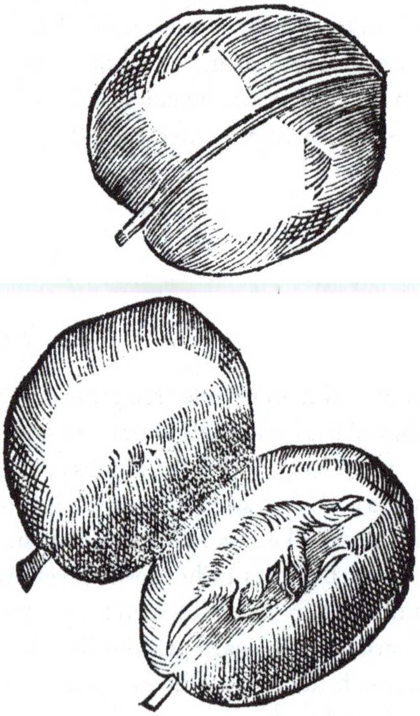

The Temperature and Vertues.

The Dragon's blood which is thought to proceed from this tree has an astringent faculty, and is with good success used in the overmuch flowing of the courses, in fluxes, dysenteries, spitting of blood, fastening loose teeth, and such other effects which require astriction. Smiths also use it to varnish over their works, to give them a sanguine color and keep them from rust.

Of the Elder tree.

The Description.

The common Elder grows up now and then to the bigness of a mean tree, casting his boughs all about, and often times remaining a shrub. The body is almost all woody, having very little pith within, but the boughs, and especially the young ones which be jointed, are full of pith within, and have but little wood without. The bark of the body and great arms is rugged and full of chinks, and of an ill-favoured wan colour as of ashes. That of the boughs is not very smooth, but in colour almost like; and that is the outward bark, for there is another under it nearer to the wood, of colour green. The substance of the wood is sound, somewhat yellow, and that may be easily cleft. The leaves consist of five or six particular ones fastened to one rib, like those of the Walnut tree, but every particular one is lesser, nicked in the edges, and of a rank and stinking smell. The flowers grow on spokie rundles which be thin and scattered, of a white colour and sweet smell. After them grow up little berries, green at the first, afterwards black, whereout is pressed a purple juice, which being boiled with Allom and such like things, does serve very well for the painter's use, as also to colour vinegar. The seeds in these are a little flat and somewhat long.

Sambucus.

The common Elder tree.

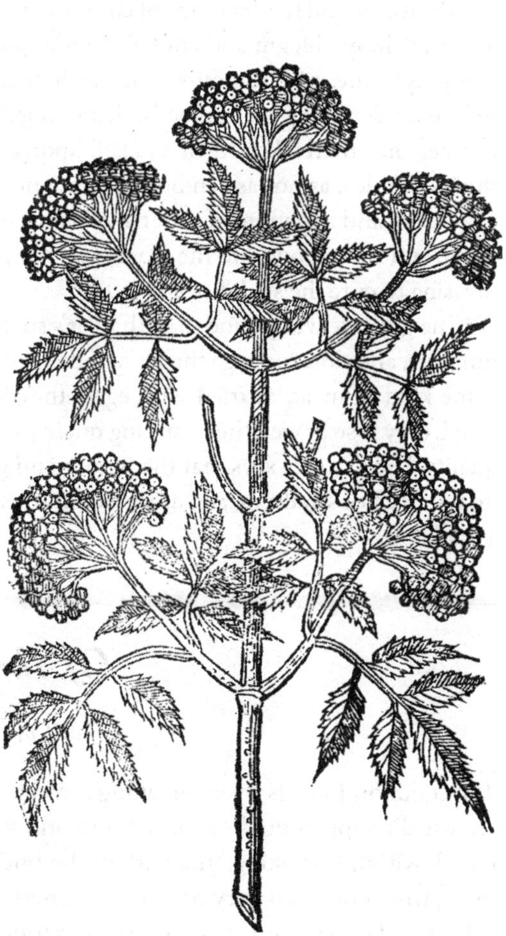

The Place and Time.

This Elder grows every where and is the common Elder. It is planted about coney burrows for the shadow of the coneys. They do flower in April and May, and their fruit is ripe in September.

The Names.

This tree is called in Latin, and of the Apothecaries, *sambucus.* In High Dutch, *holunder* and *holder;* in Italian, *sambuco;* in French, *hus;* in Spanish, *sanbugueyro;* in English, Elder and Elder tree.

The Temperature and Vertues.

Galen says that it is of a drying quality, gluing, and moderately digesting; and it has not only these qualities but others also; for the bark, leaves, first buds, flowers and fruit of the Elder do not only dry, but also heat, and have withal a purging quality, but not without trouble and hurt to the stomach.

The leaves and tender crops of common Elder, taken in some broth or potage, opens the belly, purging both slimy phlegm and cholericke humours. The middle bark is of the same nature, but stronger, and purges the said humours more violently. The leaves of Elder boiled in water until they be very soft, and when they are almost boiled enough, a little oile of sweet Almonds added thereto, or a little Linseed oile, then taken forth and laid upon a red cloth or a piece of scarlet, and applied to the hemorrhoids or piles, as hot as can be suffered, and so let to remain upon the part affected until it be somewhat cold, and applying one after another upon the diseased part, by the space of an hour or more and in the end some bound to the place, and the patient put warm abed; it has not as yet failed at the first dressing to cure the said disease.

The green leaves pounded with deer's suet or bull's tallow, are good to be laid to hot swellings and tumours, and does assuage the pain of the gout. Of like operation also are the fresh flowers mixed with some kind of meat, as fried with eggs; they likewise trouble the belly and move to the stool. Being dried, they lose as well their purging quality as their moisture, and retain the digesting and attenuating quality. Dioscorides says that the tender and green leaves of the Elder tree, with barley meal parched, do remove hot swellings, and are good for those that are burnt or scalded, and for such as be bitten with a mad dog.

Of the Elme tree.

The Description.

The common Elme is a very great high tree, the bark of the young trees and boughs of the elder, which are usually lopped or shred, is smooth and very tough, and will strip or peel from the wood a great length without breaking. The bark of the body of the old trees, as the trees grow in bigness, tears or rents, which makes it very rough. The innermost wood of the tree is of a reddish yellow or brownish colour, and curled, and after it is dry, very tough, hard to cleave or rent, whereof naves of carts are most commonly made. The wood next the bark is white. Before the leaves come forth, the flowers appear, about the end of March, which grow on the twigs or branches, closely compacted or thrust together, and are like to the chives growing in the middle of most flowers, of a reddish colour; after which come flat seed, more long than broad, not much unlike the garden Arach seed in form and bigness, and does for the most part fall away before or shortly after the leaves spring forth, and some hang on a great part of the Summer. The leaves grow on the twigs, of a dark green colour; the middle size whereof are two inches broad and three inches long; some are longer and broader, some narrower and shorter, rough or harsh in handling on both sides, nicked or indented about the edges, and many times crumpled, having a nerve in the middle, and many smaller nerves growing from him. The leaf on one side of the nerve is always longer than on the other. On these leaves often times grow blisters or small bladders, in which at the Spring are little worms, about the bigness of bed-fleas.

ulmus vulgatiss. folio lato scabro.
The common Elme tree.

The Place and Time.

The Elme grows plentifully in all places of England. The seeds of the Elme shows it self first, and before the leaves; it falls in the end of April, at what time the leaves begin to spring.

The Names.

It is called in Latin, *ulmus;* in French, *orme;* in Italian, *olmo;* in Spanish, *uimo;* in English, Elme tree.

The Temperature and Vertues.

The leaves and bark of the Elme be moderately hot, with an evident cleansing faculty. They have in the chewing a certain clammy and gluing quality. The leaves of Elme glue and heal up green wounds, so does the bark wrapped and swaddled about the wound like a band. The leaves being stamped with vinegar do take away scurf. Dioscorides writes that one ounce weight of the thicker bark drunk with wine or water, purges phlegm. The decoction of Elme leaves, as also of the bark or root, heals broken bones very speedily, if they be fomented or bathed therewith. The liquor that is found in the blisters does beautify the face, and scours away all spots, freckles, pimples, spreading tetters and such like.

Of the Fig tree.

The Description.

The garden Fig tree becomes a tree of a mean stature, having many branches full of white pith within, and large leaves of a dark green colour, divided into sundry sections or divisions. The fruit comes out of the branches without any flower at all that ever I could perceive, which fruit is in shape like unto Pears, of colour either whitish, or somewhat red or of a deep blew, full of small grains within, of a sweet and pleasant taste; which being broken before it be ripe, does yield most white milk, like unto the kinds of Spurge, and the leaves also being broken do yield the like liquor; but when the Figs be ripe, the juice thereof is like honey.

ficus.
The Fig tree.

The Place and Time.

The Fig trees do grow plentifully in Spaine and Italy, and many other countries, as in England; where they bear fruit, but it never comes to kindly maturity, except the tree be planted under a hot wall, whereto neither north nor north-east winds can come.

In England the Fig trees put not forth their leaves until the end of May, where often times the fruit comes before the leaves appear.

The Names.

The Fig tree is called in Latin, *ficus* and *ficus sativa;* in French, *figuier;* in Italian, *fico;* in Spanish, *higuera;* in English, Fig tree. The fruit is named in Latin, *ficus;* and the unripe fruit in Latin, *grossus;* that which is dried is called in Latin, *carica.* In French, *figues;* in Italian, *fichi;* in Spanish, *higos;* in English, Fig.

The Temperature and Vertues.

The green Figs new gathered are somewhat warm and moist. The dry and ripe Figs are hot almost in the third degree, and withal sharp and biting. The leaves also have some sharpness, with an opening power, but not so strong as the juice.

The dry Figs do nourish better than the green or new Figs; notwithstanding they ingender not very good blood, for such people as do feed much thereon do become lowsie.

Figs be good for the throat and lungs, they mitigate the cough, and are good for them that be short winded. They ripen phlegm, causing the same to be easily spit out, especially when they be sodden with Hyssop and the decoction drunk. Figs stamped with salt, rue, and the kernels of nuts, withstand all poison and corruption of the air. The King of Pontus, called Mithridates, used this preservative against all venom and poison.

Figs stamped and made into the form of a plaister with wheat meal, the powder of Fœnugreek, Lineseed, and the roots of marish Mallowes, applied warm, do soften and ripen impostumes, phlegmons, and all hot and angry swellings and tumours behind the ears. If you add thereto the roots of Lillies, it ripens and breaks Venerious impostumes that come in the flank, which impostume is called bubo, by reason of his lurking in such secret places. In plain English terms, they are called botches.

The leaves of the Fig tree do waste and consume the King's Evill, or swelling kernels in the throat, and do mollify, waste and consume all other tumours, being finely pounded and laid thereon; but after my practise, being boiled with the roots of marish Mallows until they be soft, and so incorporated together, and applied in form of a plaister.

The milky juice either of the Figs or leaves is good against all roughness of the skin, lepries, spreading sores, tetters, small pockes, measles, wheales, freckles, lentiles, and all other spots, scurviness and deformity of the body and face, being mixed with barley meal and applied. It does also take away warts and such like excrescences, if it be mingled with some fatty or greasy thing.

The milk does also cure the tooth ache, if a little lint or cotton be wet therein, and put into the hollowness of the tooth. The milk thereof put into the wound proceeding of the biting of a mad dog, or any other venomous beast, preserves the parts adjoining, takes away the pain presently, and cures the hurt. Dioscorides says that the white liquor of the Fig tree and juice of the leaves do curdle milk as rennet does, and dissolves the milk that is cluttered in the stomach, as does vinegar. It brings down the menses, if it be applied with the yolke of an egg, or with yellow wax.

Of the arched Indian Fig tree.

The Description.

This rare and admirable tree is very great, straight, and covered with a yellowish bark tending to tawny. The boughs and branches are many, very long, tough, and flexible, growing very long in short space, as do the twigs of Oziars, and those so long and weak that the ends thereof hang down and touch the ground, where they take root and grow in such sort that those twigs do become great trees. And these being grown up into the like greatness, do cast their branches or twiggy tendrels unto the earth, where they likewise take hold and root, by means whereof it comes to pass that of one tree is made a great wood, which the Indians do use for couverture against the extreme heat of the Sun, wherewith they are grievously vexed.

Some likewise use them for pleasure, cutting down by a direct line a long walk, or as it were a vault, through the thickest part, from which also they cut certain loop-holes or windows in some places, to receive thereby the fresh cool air that enters thereat, as also for light. From which vault or close walk does rebound such an admirable echo or answering voice, if one of them speak unto another aloud, that it does answer again four or five times, and that so plainly that it cannot be known from the voice it self.

The first or mother of this wood is hard to be known from the children, but by the greatness of the body, which three men can scarcely fathom about. Upon the branches whereof grow leaves hard and wrinkled, in shape like those of the Quince tree, green above and of a whitish hoary colour underneath, whereupon the Elephants delight to feed; among which leaves come forth the fruit, of the bigness of a man's thumb, in shape like a small Fig, but of a sanguine or bloudy colour and of a sweet taste, but not so pleasant as the Figs of Spaine; notwithstanding they are good to be eaten, and withal very wholesome.

The Place and Time.

This wondrous tree grows in diverse places of the East Indies, especially near unto Goa, and also in Malaca. It is a stranger in most parts of the world. This tree keeps his leaves green winter and summer.

arbor ex goa, siue indica.
The arched Indian Fig tree.

The Names.

This tree is called of those that have travelled, *ficus indica,* the Indian Fig; and *arbor goa,* of the place where it grows in greatest plenty. We may call it in English, the arched Fig tree.

The Temperature and Vertues.

We have nothing to write of the temperature or vertues of this tree, of our own knowledge. Neither have we received from others more, than that the fruit hereof is generally eaten, and that without any hurt at all, but rather good and also nourishing.

Of the Firre or Deale tree.

The Description.

The Firre tree grows very high and great, having his leaves ever green; his trunk or body smooth, even and straight, without joints or knots until it has gotten branches; which are many and very spare, beset with leaves, not much unlike the leaves of the Ewe tree, but smaller. Among which come forth flowers upon the taller trees, growing at the bottoms of the leaves like little catkins, as you may see them exprest in a branch apart by themselves. The fruit is like unto the Pine Apple, but smaller and narrower, hanging down as the Pine Apple. The timber hereof excels all other timbers for the making of ships, posts, rails, deale boards, and sundry other purposes.

There is another kind of Firre tree, which is likewise a very high and tall tree, and higher than the Pine. The body of it is straight without knots below, waxing smaller and smaller even to the very top; about which it sends forth boughs, four together out of one and the self same part of the body, placed one against another in manner of a cross, growing forth of the four sides of the body and observing the same order even to the very top. Out of these boughs grow others also, but by two and two, one placed right against another, out of the sides; which bend downwards when the others bear upwards.

The leaves compass the boughs round about, and the branches thereof. They be long, round, and blunt pointed; narrower and much whiter than those of the Pitch tree, that is to say of a light green, and in the manner of a white colour. The cones or clogs be long, and longer than any others of the cone trees; they consist of a multitude of soft scales, they hang down from the end of the twigs and do not easily fall down, but remain on the tree a very long time. The kernels in these are small, not greater than

abies.

The Firre tree.

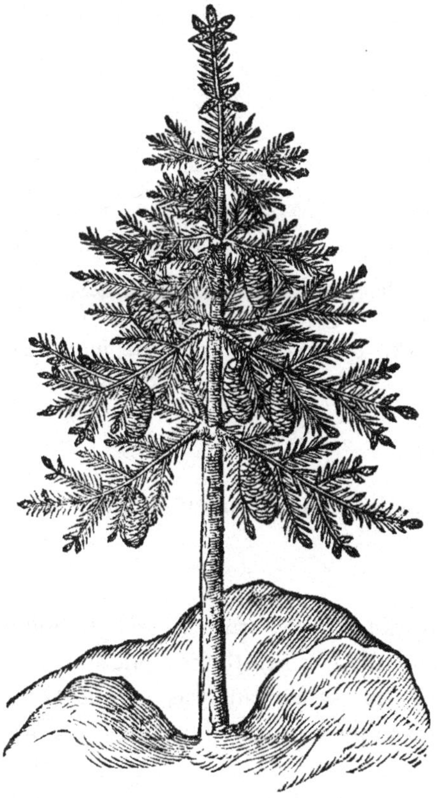

abies mas.

The male Firre tree.

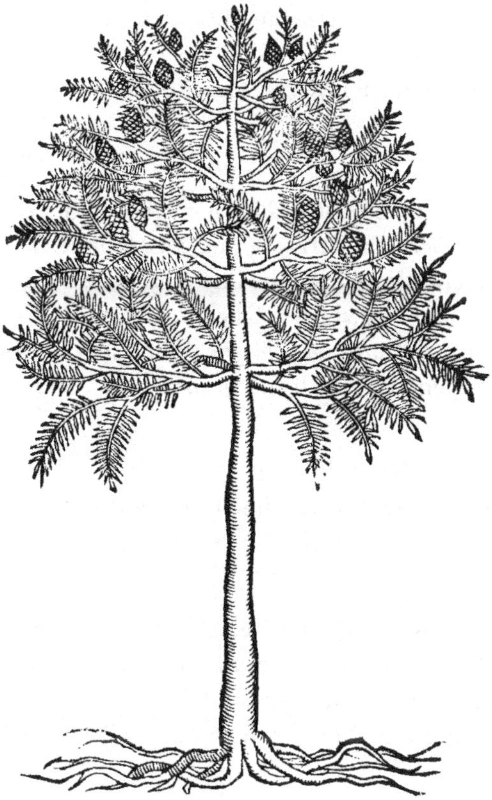

the kernels of the Cherrie stone, with a thin skin growing on the one side, very like almost to the wings of Bees or great Flies. The timber or substance of the wood is white, and clad with many coats, like the head of an Onion.

The Place and Time.

The Firre trees grow upon high mountains, in many woods of Germany and Bohemia, in which it continues always green. It is found also on the hills in Italy, France, & other countries; it comes down oftentimes into the valleys. They are found likewise in Pruse, Pomerania, Lieseland, Russia, & especially in Norway, where I have seen the goodliest trees in the world of this kind, growing upon the rockie and craggie mountains, almost without any earth about them, or any other thing, saving a little mosse about the roots, which thrust them selves here and there into the chinks and crannies of the rocks, and therefore are easily cast down with any extreme gale of wind.

I have seen these trees growing in Cheshire, Staffordshire, and Lancashire, where they grew in great plenty, as is reported, before Noah's flood. But then being overturned and overwhelmed have lain since in the mosses and watery moorish grounds very fresh, and found until this day, & so full of a resinous substance that they burn like a Torch or Linke; and the inhabitants of those countries do call it Firwood, and Fire-wood unto this day.

Out of this tree issues the rosin called *thus*; in English, Frankincense. But from the young Fir trees proceeds an excellent clear and liquid Rosin, in taste like to the peelings or outward rinde of the Pomecitron.

The time of the Fir tree agrees with the Pine trees.

The Names.

The tree is called in Latine, *abies;* in High Dutch, *weisz thannen* and *weisz thannen baum;* in Low Dutch, *witte dennen boom* or *abel boom,* and *mast boom.* In Italian, *abete;* in Spanish, *abeto,* in English, Firre-tree, Mast-tree, and Deale-tree. The first is called in French, *du Sap* or *Sapin;* the other is *suiffe.*

The liquid rosin which is taken forth of the bark of the young Firre-trees, is called in Latine, *lachryma abietis;* and *lachryma abiegna.* In the shops of Germany, as also of England, *terebinthina veneta* or Venice Turpentine; in Italian, *lagrimo.*

The Temperature.

The bark, fruit and gums of the Fir tree, are of the nature of the Pitch tree and his gums.

The Vertues.

The liquid Rosin of the Fir tree called Turpentine, loosens the belly, drives forth hot cholerick humours, cleanses and mundifies the kidnies, provokes urine, and drives forth the stone and gravel.

The same taken with Sugar and the powder of Nutmegs, cures the strangurie, stays the Gonorrhœa or the involuntary issue of man's nature, called the running of the reins and the white flux in women.

It is very profitable for all green and fresh wounds, especially the wounds of the head, for it heals and cleanses mightily, especially if it be washed in Plaintaine water and afterward in Rose water, the yolke of an egg put thereto, with the powders of *olibanum* and Masticke finely searced, adding thereto a little Saffron.

Of Fisticke Nuts

The Description.

The tree which bears Fisticke Nuts is like to the Turpentine tree. The leaves hereof be greater than those of the Masticke tree, but set after the same manner, and in like order that they are; being of a faint yellow colour out of a green.

The fruit or Nuts do hang by their stalks in clusters, being greater than the Nuts of Pine Apples and much lesser than Almonds. The husk without is of a grayish colour, sometimes reddish; the shell brickle and white, the substance of the kernel green; the taste sweet, pleasant to be eaten, and something sweet of smell.

The Place and Time.

Fisticke Nuts grow in Persia, Arabia, Syria, and in India. Now they are made free Denizons in Italy, as in Naples and in other Provinces there.

This tree does flower in May, and the fruit is ripe in September.

The Names.

This Nut is called in Latine, *pistacion, bistacion,* or *phistacion.* The Apothecaries, *fistici;* the Spaniards, *alhocigos* and *fisticos;* in Italian, *pistacchi;* in English, Fisticke Nut.

The Temperature and Vertues.

The kernels of the Fisticke Nuts are oftentimes eaten as be those of the Pine Apples; they be of temperature hot and moist; they are not so easily concocted, but much easier than common nuts. The juice is good, yet somewhat thick. They yield to the body no small nourishment; they nourish bodies that are consumed; they recover strength.

They are good for those that have the phthisicke, or rotting away of the lungs.

They concoct, ripen, and cleanse forth raw humours that cleave to the lights and chest.

They open the stoppings of the liver, and be good for the infirmities of the kidneys; they also remove out of the kidneys stones and gravel, and assuage their pain. They are also good for ulcers.

The kernels of Fisticke nuts condited, or made into comfits, with sugar and eaten, do procure bodily lust, unstop the lungs and the breast, are good against the shortness of breath, and are an excellent preservative medicine being ministered in wine against the bitings of all manner of wilde beasts.

Of the Frankincense tree.

The Description.

The tree from which Frankincense flows is but low, and has leaves like the Mastick tree, yet some are of opinion that the leaf is like the leaf of a Pear tree, and of a grassy colour. The rind is like to that of the Bay tree.

There are two kinds hereof: the one grows in mountains and rocky places, the other in the plain; but those in the plains are much worse than those of the mountains. The gum hereof is also blacker, fitter to mingle with pitch and such other stuff to trim ships, than for other uses.

arbor thurifera.
The Frankincenſe tree.

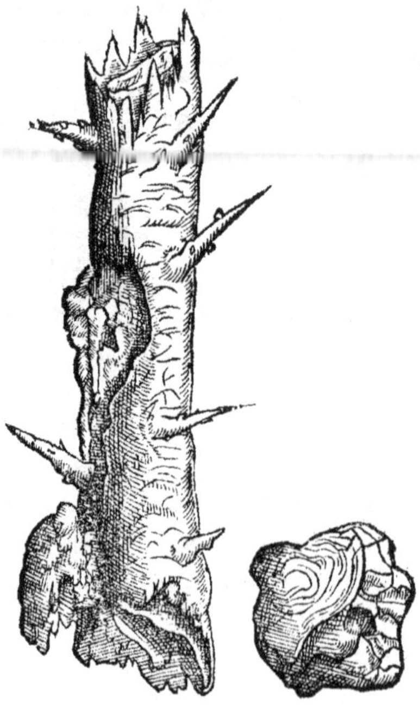

Theuet in his Cosmographie says that the Frankincense tree does resemble a gummy or rosiny Pine tree, which yields a certain juice that in time grows hard, and is called thus, Frankincense, in whom is found sometime certain small grains like unto gravel, which they call the Manna of Frankincense.

Of this there is in Arabia two other sorts, the one, the gum whereof is gathered in the Dog days when the Sun is in Leo, which is white, pure, clear, and shining. Pena writes that he has seen the clear Frankincense called *limpidum*, and yielding a very sweet smell when it is burnt, but the leaf has been seldom seen. The other is gathered in the Spring, which is reddish and worse than the other in price or value, because it is not so well concocted in the heat of the Sun. The Arabians wound this tree with a knife, that the liquor may flow out more abundantly, whereof some trees yield threescore pounds of Frankincense.

The Place.

Dioscorides says it grows in Arabia, and especially in that quarter which is called Thurisera, the best in that country is called *stagonias*, and is round; and if it be broken, is fat within; and when it is burned does quickly yield smell. Next to it in goodness is that which grows in Smilo, lesser than the other and more yellow.

The Names.

It is called in Latin, *thus;* in Italian, *incenso;* in Spanish, *enceso;* in French, *enceus;* in English, Frankincense and Incence.

The rosin carries the same name, but in shops it is called *olibanum.*

The Temperature and Vertues.

It has, as Disocorides says, a power to heat and bind. It drives away the dimness of the eyesight, fills up hollow ulcers, it closes raw wounds, stays all corruptions of blood, although it fall from the head.

The fume or smoke of it has a more drier and hotter quality than the Frankincense it self, being dry in the third degree. Dioscorides says that if it be drunk by a man in health, it drives him into a frenzie, but there are few Greeks of his mind.

Avisen reports that it does help and strengthen the wit and understanding, but the often taking of it will breed the head-ache, and if too much of it be drunk with wine, it kills.

Of Furzes, Gorsse, Whin, or prickley Broome.

The Kindes.

There be diverse sorts of prickly Broome, called in our English tongue by sundry names, according to the speech of the country people where they do grow. In some places, Furzes; in others, Whins, Gorsse, and of some, Prickly Broome.

The Description.

The Furze bush is a plant altogether a Thorne, fully armed with most sharp prickles, without any leaves at all except in the spring, and those very few and little, and quickly falling away. It is a bushy shrub, often rising up with many woody branches to the height of four or five cubits, or higher, according to the nature and soil where they grow. The greatest and highest that I did ever see do grow about Excester in the West parts of England, where the great stalks are dearly bought for the better sort of people, and the small thorny sprays for the poorer sort.

From these thorny branches grow little flowers like those of Broome, and of a yellow colour, which in hot Regions under the extreme heat of the Sun are of a very perfect red colour. In the colder countries of the East, as Danzicke, Brunswicke, and Poland, there is not any branch hereof growing, except some few plants and seeds which my self have sent to Elbing, otherwise called Meluin, where they are most curiously kept in their fairest gardens, as also our common Broome, the which I have sent thither likewise, being first desired by diverse earnest letters. The cods follow the flowers, which the Graver has omitted, as a German who had never seen the plant it self, but framed the figure by hear-say. The root is strong, tough, and woody.

We have in our barren grounds of the North parts of England another kinde of Furze, bringing forth the like prickley thornes that the other have; the only difference consists in the colour of the flowers, for the others bring forth yellow flowers, and those of this plant are as white as snow.

The Place and Time.

The common sort hereof are very well known to grow in pastures and fields in most places of England. They flower from the beginning of May to the end of September.

The Names.

Furze is commonly called *genista spinosa*. In High Dutch, *gaspeldozen;* in English, Furze, Furzen bushes, Whinne, Gorsse, and Thorne-Broome.

genista spinosa vulgaris.
Great Furze bush.

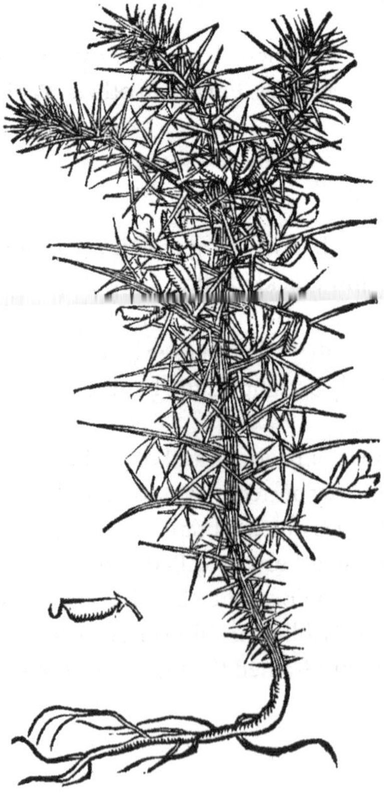

genista spinosa minor.
The small Furze bush.

The Temperature and Vertues.

There is nothing written in Theophrastus concerning the faculties of Furze; notwithstanding, the later Writers do agree that it is hot and dry of complexion. The seeds are used in medicines against the stone, and staying of the laske.

Of the Gall tree.

The Description.

The Gall tree grows up to a sufficient height, having a very fair trunk or body, whereon are placed long twiggy branches which bring forth very fair leaves, broad, and nicked in the edges like the teeth of a saw; among which come Acornes (although the figure expresses not the same) like those of the Oke, and likewise a woody excrescence, which we name the Gall, having certain small eminences or bunches on the out side, growing for the most part upon the slender branches without stalks, and sometimes they grow at the ends thereof; which by the heat of the Sun are harder, greater, and more solid in one country than another, according to the soil or climate.

galla, siue robur maius.
The great Gall tree.

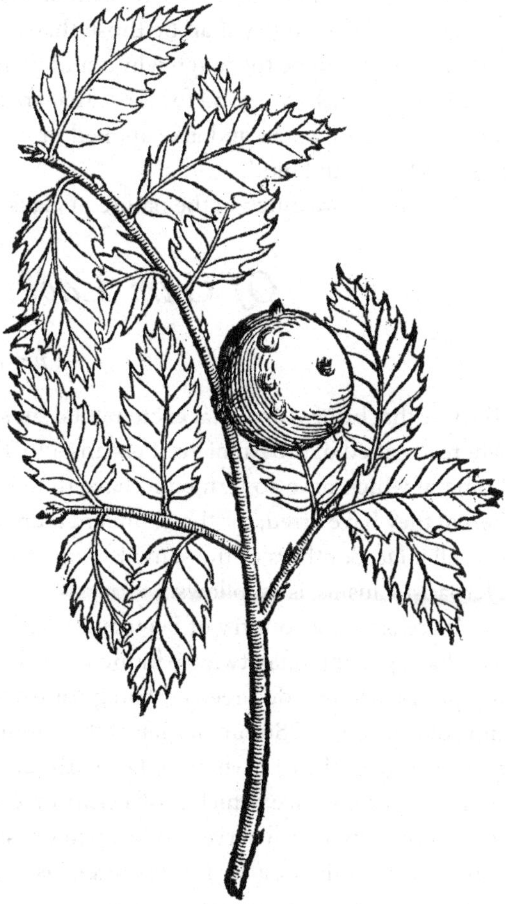

The Place and Time.

The Galls are found in Italy, Spaine, and Bohemia, and most of the hot regions. The Gall, says Pliny, appears or comes forth when the Sun comes out of the sign Gemini, and that generally in one night.

The Names.

The Gall tree is called *quercus, robur,* and *gallæ abor.* The Apothecaries and Italians keep the name *galla* for the fruit. In High Dutch, *galopssel;* in Low Dutch, *galnoten;* in French, *noix de galle;* in English, Gaules and Galls.

The Temperature and Vertues.

The Gall, as Galen writes, is dry in the third degree and cold in the second. It is a very harsh medicine, it fastens and draws together faint and slack parts, as the overgrowings of the flesh. It repels and keeps back rheums and such like fluxes, and does effectually dry up the same, especially when they have a descent into the gums, almonds of the throat, and other places of the mouth.

Galls are very profitable against the Dysentery and the Cœliacke passion, being drunk in wine, or the powder thereof strewed upon meats. They are used in dyeing and colouring of sundry things, and in making of ink.

Last of all, burnt Galls do receive a further faculty, namely to stanch blood, and are of thin parts, and of a greater virtue to dry than be those that are not burnt. They must be laid upon hot burning coals until they come to be thorough white, and then they are to be quenched in vinegar and wine.

Moreover, Galls are good for those that are troubled with the bloody flix and common laskes, being taken in wine or water, and also applied or used in meats. Finally, these are to be used as oft as need requires to dry and bind.

Oke apples are much of the nature of Galls, yet are they far inferior to them, and of lesser force.

Of Gum Lacke and his rotten tree.

The Description.

The tree that brings forth that excremental substance called *lacca,* both in the shops of Europe and elsewhere, is called of the Arabians, Persians and Turkes, *loc sumutri,* as who should say, *Lacca* of Sumatra. Some of which have so termed it, have thought that the first plenty thereof came from Sumatra, but herein they have erred, for the abundant store thereof came from Pegu, where the inhabitants thereof do call it *lac,* & others of the same Province, *trec.* The history of which tree, according to that famous Herbarist Clusius, is as follows:

There is in the country of Pegu and Malabar, a great tree, whose leaves are like them of the Plum tree, having many small twiggie branches. When the trunk or body of the tree waxes old, it rots in sundry places, wherein do breed certain great Ants or Pismires, which continually work and labour in the time of Harvest and Summer, against the penury of Winter. Such is the diligence of those Ants, or such is the nature of the tree wherein they harbour, or both, that they provide for their winter food, a lump or masse of substance, which is of a crimson colour, so beautiful and so fair, as in the whole World the like is not seen, which serves not only to physical uses, but is a perfect and costly colour for Painters, called by us, Indian Lack. The Pismires (as I said) work out this colour, by sucking the substance or matter of Lacca from the tree, as Bees do make honie and wax, by sucking the matter thereof from all herbes, trees and flowers, and the inhabitants of that country do as diligently seek for this Lacca, as we in England and other countries seek in the woods for honie. Which Lacca after they have found it, they take from the tree and dry it into a lump. Among which sometimes there come over some sticks and pieces of the tree from the wings of the Ants, which have fallen among it, as we daily see.

lacca cum suis bacillis.

Gum Lacke with his staffe or sticke.

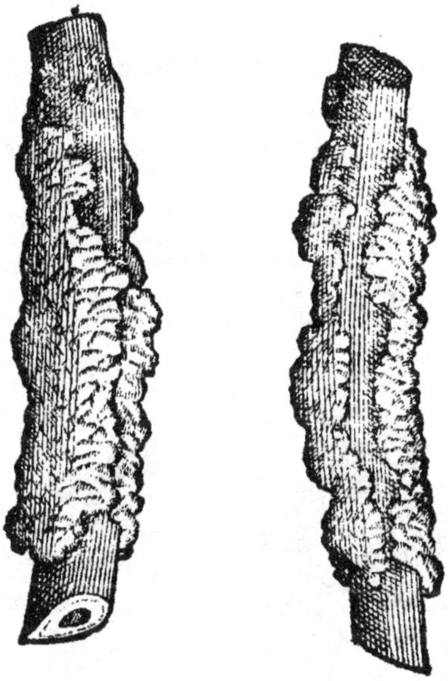

[Thomas Johnson writes: The Indian Lacke or Lake which is the rich colour used by Painters, is none of that which is used in shops, nor here figured or described by Clusius, wherefore our Author was much mistaken in that he here confounds together things so different. For this is of a resinous substance, and a faint red colour, and wholly unfit for Painters, but used alone and in composition to make the best hard sealing wax. The other seems to be an artificial thing, and is of an exquisite crimson colour, but of what it is, or how made, I have not as yet found any thing that carries any probability of truth.]

The Place and Time.

The tree which bears Lacca grows in Zeilan and Malavar, and other parts of the East Indies. Of the time we have no certain knowledge.

The Names.

Indian Lacke is called in the shops, *lacca;* in Italian, *laccetta.* Avicen calls it *luch;* Paulus and Dioscorides, as some have thought, *cancamum.* The other names are expressed in the description.

The Temperature and Vertues.

Lack or Lacca is hot in the second degree, it comforts the heart and liver, opens obstructions, expels urine, and prevails against the dropsie.

There is an artificial Lack made of the scrapings of Brasill and Saffron, which is used of Painters, and not to be used in Physicke as the other natural Lacca.

Of the Hasell tree.

The Description.

The Hasell tree grows like a shrub or small tree, parted into boughs without joints, tough and pliable. The leaves are broad, greater and fuller of wrinkles than those of the Alder tree, cut in the edges like a saw, of colour green, and on the back side more white; the bark is thin. The root is thick, strong, and growing deep. Instead of flowers hang down catkins, aglets, or blowings, slender and well compact, after which come the nuts standing in a tough cup of a green colour, and jagged at the upper end, like almost unto the beards in Roses. The shell is smooth and woody, the kernel within consists of a white, hard, and sound pulp, and is covered with a thin skin, oftentimes red, most commonly white; this kernel is sweet and pleasant unto the taste.

The Place and Time.

The Hasell trees do commonly grow in woods and in dankish untoiled places. They are also set in orchards, the nuts whereof are better and of a sweeter taste, and be most commonly red within. The catkins or aglets come forth very timely, before Winter be fully past, and fall away in March or April, so soon as the leaves come forth. The nuts be ripe in August.

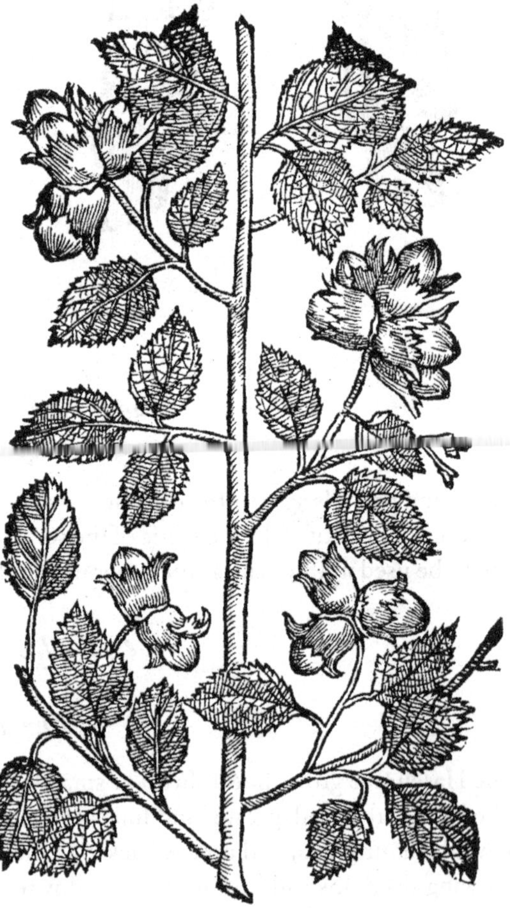

corylus sylvestris.
The wild hedge Nut.

The Names.

This shrub is called in Latin, *corylus;* in Low Dutch, *haseleer;* in English, Hasel tree and Filberd tree; but the Filberd tree is properly that which grows in gardens and orchards, and whose fruit is wholly covered over with the husk, and the shell is thinner. The nut is named in Latin, *nux pontica;* in Low Dutch, *hasel nooten;* in French, *noisettes* and *noiselles;* in English, Hasell Nut and Filberd.

The Temperature and Vertues.

Hasell nuts newly gathered and not as yet dry, contain in them a certain superfluous moisture, by reason whereof they are windy. Not only the new gathered nuts, but the dry also, be very hard of digestion, for they are of an earthy and cold essence, and of a hard and sound substance, for which cause also they very slowly pass through the belly, therefore they are troublesome and clogging to the stomach, causing head ache, especially when they be eaten in too great a quantity. The kernels of the nuts made into milk like Almonds do mightily bind the belly, and are good for the laske and the bloody flix. The same does cool exceedingly in hot fevers and burning agues. The catkins are cold and dry and likewise binding; they also stay the laske.

Of the Hawthorne, or white Thorne tree.

The Description.

The white Thorne is a great shrub, growing often times to the height of the Peare tree. The trunk or body is great, the boughs and branches hard and woody, set full of long sharp thorns. The leaves be broad, cut with deep gashes into diverse sections, smooth, and of a glistering green colour. The flowers grow upon spokie rundles, of a pleasant sweet smell, sometimes white, and often dasht over with a light wash of purple, after which comes the fruit, being round berries, green at the first and red when they be ripe; wherein is found a soft sweet pulp and certain whitish seed. The root grows deep in the ground, of a hard woody substance.

The Place and Time.

The Hawthorne grows in woods and in hedges near unto high-ways almost every where. It flowers in May, whereupon many do call the tree it self the May-bush, as a chief token of the coming in of May. The leaves come forth a little sooner. The fruit is ripe in the beginning of September, and is a food for birds in Winter.

The Names.

It is called in Italian, *bagaia;* in Spanish, *pirlitero;* in French, *aubespine;* in English, Whitethorne, Haw-thorne tree, and of some Londoners, May-bush.

The Nature and Vertues.

The fruit of the Hawthorne tree is very astringent. The haws or berries, as Dioscorides writes, do both stay the laske, the menses, and all other fluxes of blood. Some authors write that the stones, beaten to powder and given to drink, are good against the stone.

oxycanthus .

The Haw-thorne tree .

agrifolium.
The Holly tree.

Of the Holme, Holly, or Hulver tree.

The Description.

The Holly is a shrubby plant, notwithstanding it oftentimes grows to a tree of a reasonable bigness; the boughs whereof are tough and flexible, covered with a smooth and green bark. The substance of the wood is hard and sound, and blackish or yellowish within, which does also sink in the water, as does the Indian wood which is called *guaiacum.* The leaves are of a beautiful green colour, smooth and glib, like almost the Bay leaves but lesser, and cornered in the edges with sharp prickles, which notwithstanding they want or have few when the tree is old. The flowers be white and sweet of smell. The berries are round, of the bigness of a little Pease, or not much greater, of colour red, of taste unpleasant, with a white stone in the midst, which do not easily fall away but hang on the boughs a long time. The root is woody.

There is made of the smooth bark of this tree or shrub, Birdlime, which the birders and countrymen do use to take the birds with. They pull off the bark and make a ditch in the ground, specially in moist, boggy, or soggy earth, whereinto they put this bark, covering the ditch with boughs of trees, letting it remain there until it be rotten and putrefied, which will be done in the space of twelve days or thereabout; which done, they take it forth, and beat in mortars until it be come to the thickness and clamminess of Lime. Lastly, that they may clear it from pieces of bark and other filthiness, they do wash it very often; after which they add unto it a little oile of nuts, and after that, do put it up in earthen vessels.

The Place and Time.

The Holly tree grows plentifully in all countries. It grows green both Winter and Summer; the berries are ripe in September, and they do hang upon the tree a long time after.

The Names.

This tree or shrub is called in Latin, *agrifolium;* in Italian, *agrifolio* and *aguifoglio;* in Spanish, *azebo;* in French, *hous* and *housson;* in English, Holly, Hulver, and Holme.

The Temperature and Vertues.

The berries of Holly are hot and dry, and of thin parts, and waste away wind. They are good against the cholicke, for ten or twelve being inwardly taken bring away by the stoole thick and flegmaticke humours, as we have learned of them who oftentimes made trial thereof.

The Birdlime which is made of the bark hereof is no less hurtful than that of Misseltoe, for it is marvellous clammy: it glues up all the intrails, it shuts and draws together the guts and passages of the excrements, and by this means it brings destruction to man, not by any quality, but by his gluing substance.

Holly beaten to powder and drunk, is an experimented medicine against all the fluxes of the belly, as the dysentery and such like.

betulus, siue carpinus.
The Hornebeame tree.

Of the Hornebeame, or Hard beame tree.

The Description.

Betulus, or the Hornebeame tree, grows great, and very like unto the Elm, or Witch Hasell tree, having a great body; the wood or timber whereof is better for arrows and shafts, pulleys for mills, and such like devices, than Elm or Witch Hasell; for in time it waxes so hard, that the toughness and hardness of it may be rather compared unto horn than unto wood, and therefore it was called Hornebeame, or Hard beame. The leaves hereof are like the Elm, saving that they be tenderer. Among those hang certain triangled things, upon which be found knaps, in which is contained the fruit or seed. The root is strong and thick.

The Place and Time.

It grows plentifully in Northamptonshire, also in Kent by Gravesend, where it is commonly taken for a kind of Elm. This tree does spring in April, and the seed is ripe in September.

The Names.

The Hornebeame tree is called in Latin, *iuga,* yokes wherewith oxen are yoked together, which are also even at this time made thereof, and therefore it may be Englished, Yoke Elm. It is called in French, *carne;* in Italian, *carpino,* in English, Hornebeame, Hard beame, Yoke Elm, and in some places, Witch Hasell.

The Temperature and Vertues.

This tree is not used in medicine. The vertues are not expressed of the ancients, neither have we any certain experiments of our own knowledge more than has been said for the use of husbandry.

Of the Indian leaf.

tamalapatra.
The Indian leafe.

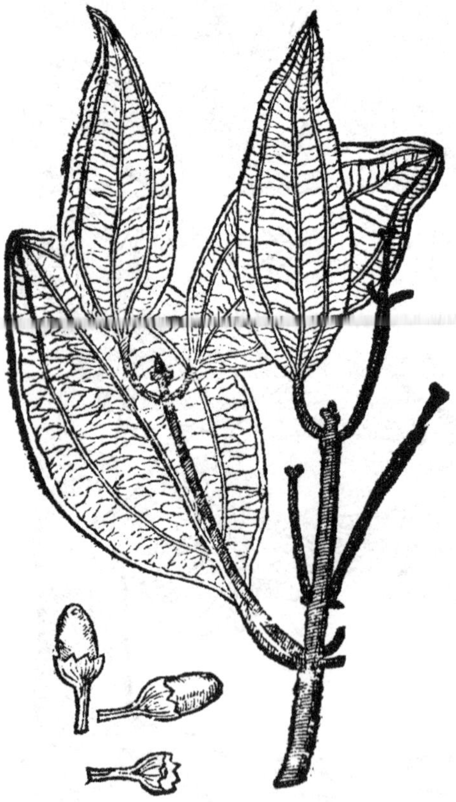

The Description.

Tamalapatra, or the Indian leaf, grows upon a great tree like the Orange tree, with like leaves also, but broader, a little sharp pointed, of a greene glistering colour, and three small ribs running through each leaf, after the manner of Ribwort, whereby it is easy to be known. It smells somewhat like unto Cloves, but not so strong as Spikenard or Mace (as some have deemed) nor yet of so subtle and quick a scent as Cinnamon.

There was sent or added unto this figure by Cortusus, a certain fruit like unto a small Acorn, with this inscription, *Fructus Canellæ,* the fruit of the Canell tree, which may be doubted of, considering the description of the forenamed tree holds generally of most to be perfect.

The Place and Time.

The Indian leaf grows not fleeting upon the water like unto *lens palustris,* as Disocorides and Pliny do set down, (though learned and painful writers) but is the leaf of a great tree, a branch whereof we have set forth unto your view, which grows in Arabia and Cambaya, far from the water side.

Of the time we have no certain knowledge, but it is supposed to be green winter and summer.

The Names.

Tamalapatra is called of the Indians, in their mother tongue, especially of the Arabians, *cadegi indi* or *ladegi indi,* that is *folium indicum,* the Indian leaf. But the Mauritanians do call it, *tembal;* the Latins and Grecians have called it, *malabathrum.*

The Temperature and Vertues.

The Indian leaf is hot and dry in the second degree, agreeing with Nardus in temperature, or as others report, with Mace. It provokes urine mightily, warms and comforts the stomacke, and helps digestion. It prevails against inflamed and watery eyes, and all other infirmities of the same. It is laid among clothes, as well to keep them from moths and other vermin, as also to give unto them a sweet smell.

Of the Indian Nut tree.

The Description.

The Grecians have not known, but the Arabians have mentioned, this Indian Nut tree, the body whereof is very great, smooth and plain, void of boughs or branches, of a great height; wherefore the Indians do wrap ropes about the body thereof, for their more ease in gathering the fruit. The timber whereof is very spongy within, but hard without, a matter fit to make their Canoos and boats of. On the top of the tree grow the leaves like those of the Date tree, but broad, and sharp at the point as thorns, whereof they use to make needles, bodkins, and such like instruments, wherewith they sew the sails of their ships, and do such like business. Among these leaves come forth clusters of flowers, like those of the Chestnut tree, which turn into great fruit of a round form, and somewhat sharp at one end; in that end next unto the tree is one hole, sometimes two, bored through. This nut or fruit is wrapped in a coverture consisting of a substance not unlike to hemp before it be beaten soft. There is also a finer and gentler stuff next unto the shell, like unto flax before it be made soft, in the middle whereof is contained a great Nut covered with a very hard shell, of a brown colour before it be polished, afterward of a black shining colour like burnished horn. Next unto the shell upon the inside, there cleaves a white cornelly substance, firm and solid, of the colour and taste of a blanched Almond. Within the cavity or hollowness thereof is contained a most delectable liquor like unto milk, and of a most pleasant taste.

The Place and Time.

This Indian nut grows in some places of Africa, and in the East Indies, and in all the islands of the West Indies, especially in Hispaniola, Cuba, and Saint John's Island, and also upon the continent by Carthagena, Nombre de Dios, and Panama, and in Virginia, otherwise called Norembega, part of the same continent, for the most part near unto the sea side, and in moist places, but it is seldom found in the uplandish countries. It grows green Winter and Summer.

The Names.

The fruit is called in Latin, *nux indica;* of the Indians, *cocus;* of the Portugals that dwell in the East Indies, *cocco,* taken from the end wherein are three holes representing the head of a Monkey; of the vulgar people, *maro,* and the fruit, *narel;* which name *narel* is common among the Persians and Arabians. It is called in Malavar, *tengamaran;* the ripe fruit, *tenga* and the green fruit, *eleri.* In Goa it is called *lanhan;* in Malaio, *triccan,* and the nut, *nihor.* The distilled liquor is called *sula* and the oile that is made thereof, *copra.*

nux indica arbor.
The Indian Nut tree.

nux indica.
The Indian Nut.

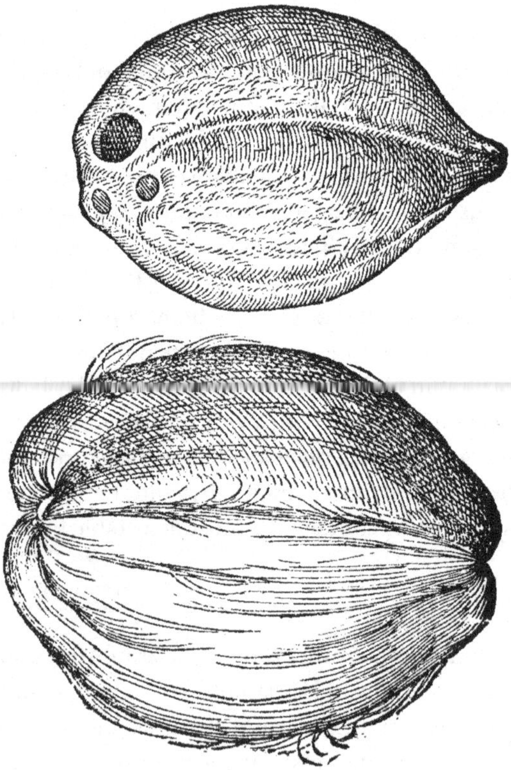

The Temperature, Vertues and Use.

It is of a mean temper betwixt hot and cold. The Indians use to cut the twigs and tender branches toward the evening, at the ends whereof they have bottle gourds, hollow canes, and such like things, fit to receive the water that drops from the branches thereof, which pleasant liquor they drink in stead of wine; from the which is drawn a strong and comfortable *aqua vitae,* which they use in time of need against all manner of sicknesses. Of the branches and boughs they make their houses; of the trunk or body of the tree, ships and boats. Of the hempon, the outward part of the fruit, they make ropes and cables; and of the finer stuff, sails for their ships. Likewise they make of the shell of the nut, cups to drink in, which we likewise use in England, garnished with silver for the same purposes. The kernel serves them for bread and meat. The milky juice does serve to cool and refresh their wearied spirits. Out of the kernel when it is stamped, is pressed a most precious oile, not only good for meat, but also for medicine, wherewith they anoint their feeble limbs after their tedious travail, by means whereof the ache and pain is mitigated, and other infirmities quite taken away proceeding of other causes.

hedera corymbosa.
Clymbing or berried Ivy.

Of Ivy.

The Description.

The greater Ivy climbs on trees, old buildings, and walls. The stalks thereof are woody, and now and then so great as it seems to become a tree; from which it sends a multitude of little boughs or branches every way, whereby as it were with arms it creeps and wanders far about. It also brings forth continually fine little roots, by which it fastens it self and cleaves wonderful hard upon trees, and upon the smoothest stone walls. The leaves are smooth, shining especially on the upper side, cornered with sharp pointed corners. The flowers are very small and mossy; after which succeed bundles of black berries, every one having a small sharp pointall.

The Place and Time.

Ivy grows commonly about walls and trees. It flourishes in Autumn. The berries are ripe after the Winter Solstice.

The Names.

Ivy is called in Latin, *hedera;* in Spanish, *yedra;* in French, *liarre.*

The Temperature and Vertues.

Ivy, as Galen says, is compounded of contrary faculties; for it has a certain binding, earthy and cold substance, and also a substance somewhat biting, which even the very taste does show to be hot. Neither is it without a third faculty, as being of a certain warm watery substance, and that is if it be green: for whilst it is in drying, this watery substance being earthy, cold, and binding, consumes away; and that which is hot and biting remains.

The leaves of Ivy fresh and green boiled in wine, do heal old ulcers, and perfectly cure those that have a venomous and malicious quality joined with them. The juice drawn or sniffed up into the nose effectually purges the head, stays the running of the ears that has been of long continuance, and heals old ulcers both in the ears and also in the nosthrils; but if it be too sharp, it is to be mixed with oile of Roses or salad oil. The leaves laid to steep in water for a day and a night's space, help sore and smarting waterish eyes, if they be bathed and washed with the water wherein they have been infused.

iuiube arabum, siue ziziphus dodonaei
The Iuiube tree.

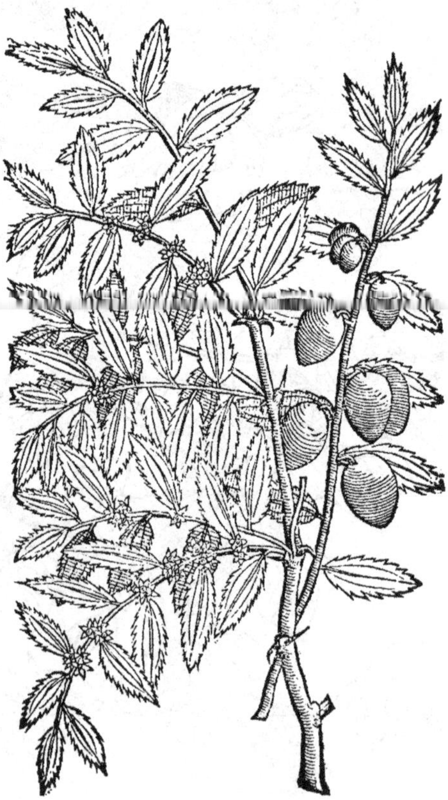

Of the Juiube tree.

The Description.

The Juiube tree is not much lesser than *ziziphus candida,* having a wreathed trunk or body, and a rough bark full of rifts or crannies, and stiff branches beset with strong and hard prickles; from whence grow out many long twigs or little stalkes, half a foot or more in length, in show like rushes, and easily bowing themselves, and very slender like the twigs of Spartum. About which come forth leaves one above another, which are somewhat long, not very great, but hard and tough like to the leaves of *peruinca* or peruinckle & among these leaves come forth pale and mossie little floures. After which succeed long red well-tasted sweet berries as big as olives (of a mean quantity) or little prunes, or small plums, wherein there are hard round stones.

The Place and Time.

There be now at this day Juiube trees growing in very many places of Italy, which in times past were newly brought thither out of Syria. It flowers in April, at which time the seeds or stones are to be set or sown for increase.

The Names.

This tree is called in Latin, *zizyphus;* and of Petrus Crescentius, *zezulus;* in English, Juiube tree. The fruit or plums are named in Latin, *zizypha* and *serica;* in shops, *Iuiubæ;* in English, Juiubes.

The Temperatures and Vertues.

Juiubes are temperate in heat and moisture. The fruit of the Juiube tree eaten is of hard digestion, and nourishes very little, but being taken in syrups, electuaries, and such like confections, it appeases and mollifies the roughnesse of the throat, the breast and lungs, and is good against the cough but exceeding good for the reines of the back, and kidneys and bladder.

Of the Juniper tree.

The Description.

The common Juniper tree grows in some parts of Kent unto the stature and bigness of a fair great tree, but most commonly it grows very low unto the ground. This tree has a thin bark or rind, which in hot regions will chop and rend it self into many crannies or pieces, out of which rifts issues a certain gum or liquor much like unto Frankincense. The leaves are very small, narrow, hard, and somewhat prickly, growing ever green along the branches, thick together; amongst which come forth round and small berries, green at the first, but afterward black declining to blewnesse, of a good savour and sweet in taste, which do wax somewhat bitter after they be dry and withered.

The Place and Time.

The common Juniper tree is found in very many places, especially in the South parts of England. The Juniper tree flowers in May; the flower whereof is nothing else but as it were a little yellowish dust or powder strewed upon the boughs. The fruit is ripe in September, and is seldom found either Winter or Summer without ripe and unripe berries, and all at one time.

The Names.

The Apothecaries keep the Latin name *juniperus.* The Arabians call it *archonas;* the Italians, *ginepro;* in Spanish, *ginebro* and *zimbro.* The French men and base Almaines call it *geneve.* In English, Juniper tree. The berries are called *grana juniperi;* in English, Juniper berries. The gum of the Juniper tree is usually called of the Apothecaries, *vernix;* in Latin, *lachryma juniperi.*

Juniperus.
The Iuniper tree.

The Temperature and Vertues.

Juniper is hot and dry and that in the third degree. The berries are also hot, but not altogether so dry. The gum is hot and dry in the first degree. The fruit of the Juniper tree does cleanse the liver and kidneys. It also makes thin clammy and gross humours. It is used in counterpoisons and other wholesome medicines. Being over largely taken, it causes gripings and gnawings in the stomach, and makes the head hot. It neither binds nor looses the belly. It provokes urine.

It is most certain that the decoction of these berries is singular good against an old cough, and against that with which children are now and then extremely troubled, called the Chin cough, in which they use to rise up raw, tough and clammy humours, that have many times blood mixed with them.

Diverse in Bohemia do take in stead of other drink, the water wherein those berries have been steeped, who live in wonderful good health. This is also drunk against poisons and pestilent fevers, and it is not unpleasant in the drinking; when the first water is almost spent, the vessel is again filled up with fresh.

The smoke of the leaves and wood drives away serpents, and all infection and corruption of the air, which bring the plague or such like contagious diseases. The juice of the leaves is laid on with wine, and also drunk, against the bitings of the viper.

The ashes of the burned bark, being applied with water, take away scurf and filth of the skin. The fume and smoke of the gum do stay phlegmaticke humours that distil out of the head, and stop the rheum. The gum does stay raw and phlegmaticke humours that stick in the stomach and guts, if it be inwardly taken, and also drunk.

It kills all manner of worms in the belly; it stays the menses and hemorrhoids; it is commended also against the spitting of blood. It dries hollow ulcers and fills them with flesh, if it be cast thereon. Being mixed with oil of Roses, it heals chops in the hands and feet.

There is made of this and of oile of Lineseed, mixed together, a liquor called Varnish, which is used to beautify pictures and painted tables with, and to make iron glitter and to defend it from the rust.

Of the Larch tree.

larix cum agarico suo .
The Larch tree with his Agarick .

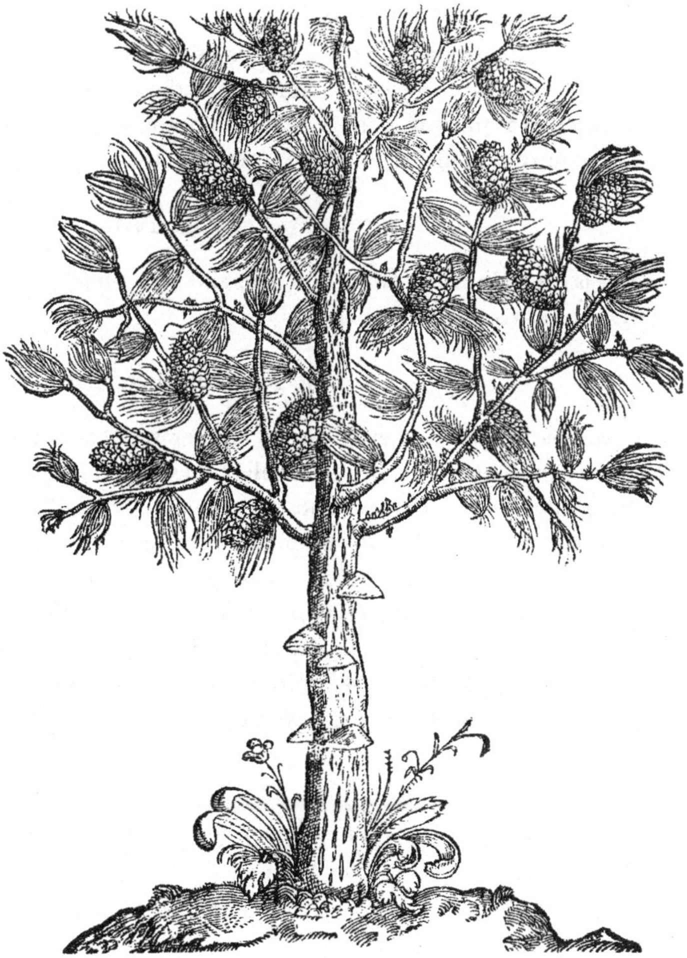

The Description.

The Larch is a tree of no small height, with a body growing straight up; whereof in the nether part beneath the boughs is thick, rugged, and full of chinks; which being cut in sunder is red within, and in the other part above, smooth, slippery, something white without. It brings forth many boughs divided into other lesser branches, which be tough and pliable. The leaves are small, and cut into many jags, growing in clusters thick together like tassels, which fall away at the approach of Winter. The flowers, or rather the first showing of the cones or fruit, be round, and grow out of the tenderest boughs, being at the length of a brave red purple colour. The cones be small, and like almost in bigness to those of the Cypress tree, but longer, and made up of a multitude of thin scales like leaves; under which lie small seeds, having a thin film growing on them very like to the wings of bees and wasps. The substance of the wood is very hard, of colour (especially that in the midst) somewhat red, and very profitable for works of long continuance.

It is not true that the wood of the Larch tree cannot be set on fire, as Vitruvius reports of the castle made of Larch wood, which Caesar besieged; for it burns in chimneys and is turned into coles, which are very profitable for Smiths, as Matthiolus writes.

There is gathered of the Larch tree a liquid Rosin, which issues not forth of it self, but runs out when the tree has been bored even to the heart with a great and long auger and wimble.

There grows also upon the Larch tree a kind of Mushrum or excrescence, not such as is upon other trees, but whiter, softer, more loose and spongie, and good for medicine, which bears the name Agarcike.

The Place and Time.

The Larch tree grows not in Greece, nor in Macedon, but chiefly upon the Alps of Italy not far from Trent, hard by the rivers Benacus and Padus; and also in other places in the same mountains. It is likewise found on the hills in Moravia.

Of all the cone trees, only the Larch is found to be without leaves in Winter. In the Spring grow fresh leaves out of the same knobs from which the former did fall. The cones are to be gathered before Winter, so soon as the leaves are gone; but after the scales are loosed and opened, the seeds drop away. The Rosin must be gathered in the Summer months.

The Names.

This tree is called in Latin, *larix;* in Italian and Spanish *larice;* in French, *melese;* in English, Larch tree and of some, Larix tree.

The Temperature and Vertues.

The leaves, bark, fruit and kernel, are of temperature like unto the Pine, but not so strong. The Larch Rosin is of a moister temperature than all the rest of the Rosins, and is withal without sharpness or biting, and is fitly mixed with medicines which perfectly cure ulcers and green wounds.

Agaricke is hot in the first degree and dry in the second. It cuts, makes thin, cleanses, takes away obstructions or stoppings of the intrailes, and purges also by stool. It cures the yellow jaundice proceeding of obstructions, and is a sure remedy for cold shakings which are caused of thick and cold humours. The same being inwardly taken and outwardly applied, is good for those that are bit of venomous beasts which hurt with their cold poison. It is good against the shortness of breath called *asthma,* and the inveterate cough of the lungs. It comforts the weak and feeble stomach, causes good digestion, and is good against worms.

Of the Line or Linden tree.

The Description.

The female Line or Linden tree waxes very great and thick, spreading forth his branches wide and far abroad; being a tree which yields a most pleasant shadow, under and within whose boughs may be made brave Summer houses and banqueting arbours, because the more that it is surcharged with weight of timber and such like, the better it does flourish. The bark is brownish, very smooth, and plain on the outside, but that which is next to the timber is white, moist, and tough, serving very well for ropes, traces, and halters. The timber is whitish, plain and without knots, yea very soft and gentle in the cutting or handling. Better gunpowder is made of the coles of this wood than of willow coles. The leaves are green, smooth, shining and large, somewhat snipt or toothed about the edges. The flowers are little, whitish, of a good savour, and very many in number, growing clustering together from out of the middle of the leaf; out of which proceeds a small, whitish, long narrow leaf. After the flowers succeed cornered sharp pointed nuts, of the bigness of Hasell nuts. This tree seems to be a kind of Elm, and the people of Essex about Henningham (whereas great plenty grows by the way sides) do call it broad leafed Elm.

The male *tilia* or Line tree grows also very great and thick, spreading it self far abroad like the other Linden tree. His bark is very tough and pliant, and serves to make cords and halters of. The timber of this tree is much harder, more knotty, and more yellow than the timber of the other, not much differing from the timber of the Elm tree. The leaves hereof are not much unlike Ivy leaves, not very green, somewhat snipt about the edges; from the middle whereof come forth clusters of little white flowers like the former, which being faded, there succeed small round pellets, growing clustering together like Ivy berries, within which is contained a little round blackish seed, which falls out when the berry is ripe.

tilia fœmina.
The female Line tree.

tilia mas.
The male Line tree.

The Place and Time.

The female Linden tree grows in some woods in Northamptonshire; also near Colchester, and in many places alongst the high way leading from London to Henningham in the county of Essex. The male Linden tree grows in my Lord Treasurer's garden at the Strand, and in sundry other places, as at Barnelmes, and in a garden at Saint Katherine's near London. These trees flower in May and their fruit is ripe in August.

The Names.

The Linden tree is called in Latin, *tilia;* in High Dutch, *linden* and *lindenbaum;* in Low Dutch, *linde* and *lindenboom;* of the Italians, *tilia;* of the Spaniards, *teia;* in French, *tilet* and *tilieul;* in English, Linden tree and Line tree.

The Temperature and Vertues.

The bark and leaves of the Linden or Line tree are of a temperate heat, somewhat drying and astringent. The leaves of *tilia,* boiled in Smithes water with a piece of Allom and a little honey, cure the sores in children's mouths. The leaves boiled until they be tender, and pounded very small with hog's grease, and the powder of fenugreeke and Line seed, take away hot swellings and bring impostumes to maturation, being applied thereto very hot. The flowers are commended by diverse against pain of the head proceeding of a cold cause; against dizziness, the apoplexy, and also the falling sickness, and not only the flowers but the distilled water thereof. The leaves of the Linden, says Theophrastus, are very sweet, and be a fodder for most kind of cattle. The fruit can be eaten of none.

Of the Maple tree.

The Description.

The great Maple is a beautiful and high tree, with a bark of a mean smoothness. The substance of the wood is tender and easy to work on. It sends forth on every side very many goodly boughs and branches, which make an excellent shadow against the heat of the sun; upon which are great broad and cornered leaves, much like those of the Vine, hanging by long reddish stalks. The flowers hang by clusters, of a whitish green colour; after them come up long fruit fastened together by couples, one right against another, with kernels bumping out near to the place in which they are combined; in all the other parts flat and thin like unto parchment, or resembling the innermost wings of grasshoppers. The kernels be white and little.

There is a small Maple which does oftentimes come to the bigness of a tree, but most commonly it grows low after the manner of a shrub. The bark of the young shoots hereof is likewise smooth; the substance of the wood is white, and easy to be wrought on. The leaves are cornered like those of the former, slippery, and fastened with a reddish stalk, but much lesser. The flowers be as those of the former, green, yet not growing in clusters but on spoked rundles. The fruit stands by two and two upon a stem or foot-stalk.

The Place and Time.

The small or hedge Maple grows almost every where in hedges and low woods. The great Maple is a stranger in England, only it grows in the walks and places of pleasure of noble men, where it especially is planted for the shadow sake, and under the name of Sycomore tree. These trees flower about the end of March, and their fruit is ripe in September.

The Names.

This tree is called in Latin, *acer;* in English, Maple or Maple tree. The great Maple is called in High Dutch, *ahozne* and *waldescherne;* by the French men, *grand erable.* This is now commonly (yet not rightly) called the Sycomore tree; and seeing we will have it so, I think it not unfit to call it the bastard Sycomore. The other is called in Latin, *acer minor;* in English, small Maple and common Maple.

acer maius.

The great Maple.

The Temperature and Vertues.

What use the Maple has in medicine we find nothing written of the Grecians, but Pliny affirms that the root pounded and applied, is a singular remedy for the pain of the liver.

Of the Lentiske, or Masticke tree.

The Description.

The Mastick tree grows commonly like a shrub, without any great body, rising up with many springs and shoots like the Hasell; and oftentimes it is of the height and bigness of a meane tree. The boughs thereof are tough and flexible; the bark is of a yellowish red colour, pliable likewise, and hard to be broken. There stand upon one rib, for the most part, eight leaves, set upon a middle rib much like to the leaves of Licorice, but harder, of a deep green colour, and oftentimes somewhat red in the brims, as also having divers veins running along, of a red colour, and something strong of smell.

The floures be mossy and grow in clusters upon long stems. After them come up the berries, of the bigness of Vetches, green at the first, afterwards of a purple colour, and last of all, black, fat, and oily, with a hard black stone within. The kernel whereof is white and of which also is made oile, as Dioscorides witnesses. It brings forth likewise cods beside the fruit (which may rather be termed an excrescence than a cod) writhed like a horn, in which lies at the first a liquor and afterwards when this waxes stale, little living things like unto gnats, as in the Turpentine hornes, and in the folded leaves of the Elm tree. There comes forth of the Mastick tree a rosin, but dry, called Masticke.

The Place.

The Mastick tree grows in many regions, as in Syria, Candy, Italy, Languedocke, and in most provinces of Spain, but the chiefest is in Chios, an Island in Greece, in which it is diligently and specially looked unto, and that for the Masticke sake, which is there gathered from the husbanded Masticke trees by the inhabitants every year most carefully, and is sent from thence into all parts of the world.

The Time.

The floures be in their pride in the spring time, and the berries in autumne. The Mastick must be gathered about the time when the Grapes be.

The Names.

This tree is named in Latine, *lentiscus;* in Italian, *lentisque;* in Spanish, *arcoyra;* in English, Masticke tree, and of some, Lentiske tree.

The rosin is called in Latine, *lentiscina resina* and likewise, *mastiche;* in shops, *mastix;* in Italian, *mastice;* in High and Low Dutch and French also, *mastic;* in Spanish, *almastiga, mastech,* and *almecega;* in English, Masticke.

Clusius writes that the Spaniards call the oile that is pressed out of the berries, *azeyte de mata.*

The Temperature.

The leaves, bark, and gum of the Masticke tree are of a meane and temperate heat, and are dry in the second degree, and somewhat astringent.

Lentiscus.
The Maſticke tree.

The Vertues.

The leaves and barke of the Masticke stop the laske, the bloody flixe, the spitting of blood, the pissing of blood, and all other fluxes of the blood. They are also good against the falling sickness, the falling down of the mother, and coming forth of the fundament.

The gum Masticke has the same vertue, if it be relented in wine and given to be drunke.

Masticke chewed in the mouth is good for the stomache, stays vomiting, increases appetite, comforts the braines, stays the falling down of the rheumes and watery humors, and makes a sweet breath.

The same infused in rose water, is excellent to wash the mouth withal, to fasten loose teeth, and to comfort the jaws. The same spread upon a piece of leather or velvet, and laid plaisterwise upon the temples, stays the rheume from falling into the jaws and teeth, and eases the pains thereof.

It prevails much against ulcers and wounds, being put into digestives and healing unguents.

It draws phlegme forth of the head gently and without trouble, it is also used in waters which serve to cleanse and make faire the face with. The decoction of this fills up hollow ulcers with flesh, if they be bathed therewith. It knits broken bones, stays eating ulcers, and provokes urine.

Of the Medlar tree.

The Description.

The manured Medlar tree is not great, the body whereof is writhed, the boughs hard, not easy to be broken. The leaves be longer yet narrower than those of the Apple tree, dark, green above, and somewhat whiter and hairy below. The flowers are white and great, having five leaves apiece. The fruit is small, round, and has a broad compassed navel or crown at the top. The pulp or meat is at the first white, and so harsh or choking that it cannot be eaten before it becomes soft; in which are contained five seeds or stones which be flat and hard.

mespilus sativa.
The garden Medlar.

The Place and Time.

The Medlar trees do grow in orchards, and often times in hedges among briers and brambles; being grafted in a Whitethorn it prospers wonderfully well, and brings forth fruit twice or thrice bigger than those that are not grafted at all, almost as great as little Apples. We have diverse sorts of them in our orchards. It is very late before Medlars be ripe, which is in the end of October, but the flowers come forth timely enough.

The Names.

In Latin, *mespilus;* in High Dutch, *nespelbaum*; in French, *nefflier;* in English, Medlar tree. The Apple or fruit is named in Latin likewise, *mespilum;* in High Dutch, *nespel;* in Italian, *nespolo;* in French, *neffle;* in Spanish, *nesperas;* in English, Medlar.

The Temperature and Vertues.

The Medlars are cold, dry and astringent; the leaves are of the same nature. Medlars do stop the belly, especially when they be green and hard, for after that they have been kept a while, so that they become soft and tender, they are then more fit to be eaten. Medlars are often times preserved with sugar or honey, and being so prepared they are pleasant to the taste. Moreover, they are singular good for women with child, for they strengthen the stomach and stay the loathsomeness thereof. The stones or kernels of the Medlars, made into powder and drunk, do break the stone, expel gravel, and procure urine.

Of Misseltoe or Misteltoe.

The Description.

Viscum or Misseltoe has many slender branches spread overthwart one another, and wrapped and interlaced one within another; the bark whereof is of a light green or popinjay colour. The leaves of this branching excrescence be of a brown green colour. The flowers be small and yellow; which being past, there appear small clusters of white translucent berries, which are so clear that a man may see through them, and are full of a clammy or viscous moisture, whereof the best bird-lime is made, far exceeding that which is made of Holme or Holly bark; and within this berry is a small black kernel or seed. This excrescence has not any root, neither does increase him self of his seed, as some have supposed; but it rather comes of a certain moisture and substance gathered together upon the boughs and joints of the tree, through the bark whereof this vaporous moisture proceeding, brings forth the Misseltoe. Many have diversely spoken hereof; some of the Learned have set down that it comes of the dung of a bird called a Thrush, who having fed of the seeds thereof, as eating his own bane, has voided and left his dung upon the tree, whereof was ingendered this berry, a most fit matter to make lime of to entrap and catch birds withal.

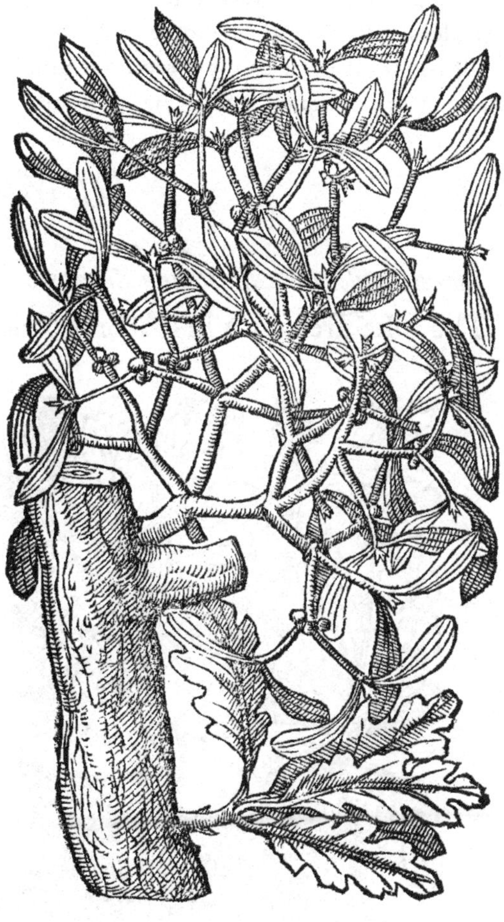

The Place and Time.

Misseltoe grows upon Okes and diverse other trees almost every where.

Misseltoe is always green as well in Winter as in Summer. The berries are ripe in Autumn. They remain all Winter through, and a food for diverse birds, as Thrushes, Black-birds, and Ring-doves.

The Names.

Misseltoe is called in Latin, *viscum;* in High Dutch, *mistell;* in Low Dutch, *marentacken;* in Italian, *vischio;* in Spanish, *liga;* in the Portugal language, *visgo;* in English, Missel and Misseltoe.

The Temperature and Vertues.

The leaves and berries of Misseltoe are hot and dry, and of subtle parts. The birdlime is hot and biting, and consists of an airy and watery substance, with some earthy quality; for according to the judgement of Galen, his acrimony overcomes his bitterness; for if it be used in outward applications, it draws humors from the deepest or most secret parts of the body, spreading and dispersing them abroad and digesting them.

It ripens swellings in the groin, hard swellings behind the ears, and other impostumes, being tempered with Rosin and a little quantity of wax. With Frankincense, it mollifies old ulcers and malicious impostumes; being boiled with unslaked lime and applied, it wastes away the hardness of the spleen. With *orpment* or *sandaraca*, it takes away foul, ill-favoured nails; being mixed with unslaked lime and wine lees, it receives greater force.

It has been most credibly reported unto me that a few of the berries of Misseltoe, bruised and strained into oile and drunken, has presently and forthwith rid a grievous and sore stitch.

Of Mosse of trees.

quercus vulgaris cum glande & musco suo.

The Oke Tree with his Acornes and Mosse.

The Description.

Tree Mosse has certain things like hairs, made up as it were of a multitude of slender leaves, now and then all to be jagged, hackt, and finely carved, twisted and interlaced one in another, which cleave fast to the barks of trees, hanging down from the bodies. One of this kind is more slender and thin, another more thick, another shorter, another longer; all of them for the most being of a whitish colour, yet oftentimes there is a certain one also which is black, but lesser and thinner. The most commendable of them all, as Pliny says, be those that are whitish, then the reddish, and lastly such as be black.

The Place and Time.

Mosse is found on the Oke tree, the white and black Poplar tree, the Olive tree, the Birch tree, the Apple tree, the Pear tree, the Pine tree, the wild Pine tree, the Pitch tree, the Fir tree, the Cedar tree, the Larch tree, and on a great sort of other trees. The best, as Dioscorides says, is that of the Cedar tree; the next, of the Poplar, in which kind the White and the sweet smelling Mosse is the chiefest; the blackish sort is of no account. Matthiolus writes that

muscus ramosus floridus.
Flouring branched Mosse.

in Italy, that Mosse is sweet which grows on the Pine tree, the Pitch tree, the Fir tree, and the Larch tree; and the sweetest, that of the Larch tree. Mosse upon the trees continues all the year long.

The Names.

Mosse is called of the Latins, *muscus;* the Arabians and some Apothecaries in other countries call it *usnea;* in High Dutch, *mosz;* in Low Dutch, *mosch;* the Frenchmen call it *la mousch;* the Italians, *musgo;* in Spanish, *musco de los arbores;* in English, Mosse, tree Mosse, or Mosse of trees.

The Temperature and Vertues.

Mosse is somewhat cold and binding, which notwithstanding is more and less according unto the nature and faculty of that tree on which it grows, and especially of his bark. For it takes unto it self and also retains a certain property of that bark, as of his breeder of which he is ingendered. Therefore the Mosse which comes of the Oke does cool and very much bind; besides his own and proper faculty, it receives also the extreme binding quality of the Oke bark it self.

Serapio says that the wine in which Mosse has been steeped certain days, brings sound sleep. It is fit to be used in compositions which serve for sweet perfumes, and that take away wearisomeness; for which things that is best of all which is most sweet of smell.

Of the Mulberrie tree.

The Description.

The common Mulberrie tree is high and full of boughs; the body whereof is many times great, the bark rugged and that of the root, yellow. The leaves are broad and sharp pointed, something hard, and nicked on the edges; in stead of flowers are blowings or catkins, which are downy. The fruit is long, made up of a number of little grains, like unto a black-Berry, but thicker, longer, and much greater, at the first green, and when it is ripe, black, yet is the juice (whereof it is full) red. The root is parted many ways.

The white Mulberrie tree grows until it be come unto a great and goodly stature, almost as big as the former. The leaves are rounder, not so sharp pointed, nor so deeply snipt about the edges, yet *morus.*

The Mulberrie tree.

sometimes sinuated or deeply cut in on the sides. The fruit is like the former, but that it is white and somewhat more tasting like wine.

The Place.

The Mulberrie trees grow plentifully in Italy and other hot regions, where they do maintain great woods and groves of them, that their silk worms may feed thereon. The Mulberrie tree is fitly set by the slip; it may also be grafted or inoculated into many trees; being grafted in a white Poplar, it brings forth white mulberries, as Beritius in his Geoponickes reports. These grow in sundry gardens in England.

The Time.

Of all the tress in the orchard, the Mulberrie does last bloom, and not before the cold weather is gone in May (therefore the old writers were wont to call it the wisest tree); at which time the silk worms do seem to revive, as having then wherewith to feed and nourish themselves, which all the Winter before do lie like small grains or seeds, or rather like the dunging of a flesh fly upon a glass, or some such thing, as knowing their proper time both to perform their duties for which they were created, and also when they may have wherewith to maintain and preserve their own bodies, unto their business aforesaid.

The berries are ripe in August and September. Hegesander affirms that the Mulberrie trees in his time did not bring forth fruit in twenty years together, and that so great a plague of the gout then reigned and raged so generally, as not only men, but boys, wenches, eunuchs and women were troubled with that disease.

The Names.

This tree is named in Latin, *morus;* in shops, *morus celsi;* in High Dutch, *maulberbaum;* in French, *meurier;* in English, Mulberrie tree. The fruit is called in Latin, *morum;* in shops, *morum celsi;* in Italian, *moro;* in French, *meure;* in Spanish, *moras;* in English, Mulberrie.

The Temperature and Vertues.

Mulberries being gathered before they be ripe, are cold and dry almost in the third degree, and do mightily bind. Being dried, they are good for the laske and bloody flix; the powder is used in meat, and is drunk with wine and water.

They stay bleedings, and also the reds. They are good against inflammations or hot swellings of the mouth and jaws, and for other inflammations newly beginning.

The ripe and new gathered Mulberries are likewise cold and be full of juice, which has the taste of wine, and is something drying and not without a binding quality; and therefore it is also mixed with medicines for the mouth, and such as help the hot swellings of the mouth and almonds of the throat, for which infirmities it is singular good.

Of the juice of the ripe berries is made a confection with sugar, called *diamorum;* that is, after the manner of a syrup, which is exceeding good for the ulcers and hot swellings of the tongue, throat, and almonds, or uvula of the throat, or any other malady arising in those parts.

The bark of the root is bitter, hot and dry, and has a scouring faculty. The decoction hereof does open the stoppings of the liver and spleen; it purges the belly and drives forth worms. The same bark being steeped in vinegar, helps the tooth-ache. Of the same effect is also the decoction of the leaves and bark, says Dioscorides, who shows that about harvest time, there issues out of the root a juice, which the next day after is found to be hard, and that the same is very good against the tooth-ache, and it wastes away *phyma* and purges the belly.

Of the Myrtle tree.

The Description.

The Myrtle is a small tree, growing to the height of a man, having many fair and pliant branches, covered with a brown bark, and set full of leaves much like unto the Laurel or Bay leaf, but thinner and smaller, somewhat resembling the leaves of Peruincle, which being bruised do yield forth a most fragrant smell, not much inferior unto the smell of Cloves. Among these leaves come forth small white flowers, in shape like the flowers of the Cherry tree, but much smaller, and of a pleasant savour, which do turn into small berries, green at the first and afterwards black.

The Place and Time.

Myrtle grows naturally upon the woody hills and fertile fields of Italy and Spaine. Where they joy to grow of themselves, they flower when the Roses do. The fruit is ripe in Autumn. In England they never bear any fruit.

myrtus laurea maxima.
The Myrtle tree.

The Names.

It is called in Latin, *myrtus;* in the Arabic tongue, *alas;* in Italian, *myrto;* in Spanish, *arrayhan;* in the Portingale language, *murta* and *murtella.* Other nations do almost keep the Latin name, as in English it is called Myrtle or Myrtle tree.

The Temperature and Vertues.

The Myrtle consists of contrary substances, a cold earthiness bearing the preheminence. It has also a certain subtle heat, as Galen says; it dries notably. They are a remedy for laskes and for the bloody flix; they quench the fiery heat of the eyes if they be laid on with parched Barley meal. It is good with a decoction made with wine, to bathe the limbs that are out of joint, and burstings that are hard to be cured, and ulcers also of the outward parts. It helps spreading tetters, scours away the dandrafe and sores of the head, makes the hairs black, and keeps them from shedding, withstands drunkenness if it be taken fasting, and prevails against poison and the bitings of any venomous beast.

Of the Nutmeg tree.

The Description.

The tree that bears the Nutmeg and the Mace is in form like to the Pear tree, but the leaves of it are like those of the Bay or Orenge tree, always green on the upper side, and more whitish underneath; among which come forth the Nut and Mace, as it were the flowers. The Nut appears first, compassed about with the Mace, as it were in the middle of a single Rose, which in process of time does wrap and enclose the Nut round on every side. After comes a husk like that of the Walnut, but of a harder substance, which incloses the Nut with his Mace as the Walnut husk does cover the Nut, which in time of ripeness does cleave of it self as the Walnut husk does, and shows his Mace, which then is of a perfect crimson colour and makes a most goodly show, especially when the tree is well laden with fruit.

[Thomas Johnson writes: I can scarce believe our author's assertion in the description, that the Nut appears first, compassed about with the Mace as it were in the middest of a single Rose, &c. But I rather think they all come forth together, the Nutmeg, Mace, the green outward husk and all, just as we see

Walnuts do, and only open them selves when they come to full maturity. In the third figure you may see expressed the whole manner of the growing of the Nutmeg, together with both the sorts of Nutmegs taken forth from their shells.]

The Place and Time.

The Nutmeg tree grows in the Indies, in an island especially called Banda, and in the islands of Molucca, and in Zeilan, though not so good as the first. The fruit is gathered in September in great aboundance, all things being common in those countries.

The Names.

The Nutmeg tree is called in Latin, *nux moschata* and *nux myristica;* in Italian, *noce moscada;* in Spanish, *nuez de escetie;* in French, *noix muscade;* of the Arabians, *leuzbane* or *gianziban.* Of the country people where they grow, *palla;* the Maces, *bunapalla.* In English, Nutmeg.

nux mscata rotunda, siue fœmina.
The round or female Nutmeg.

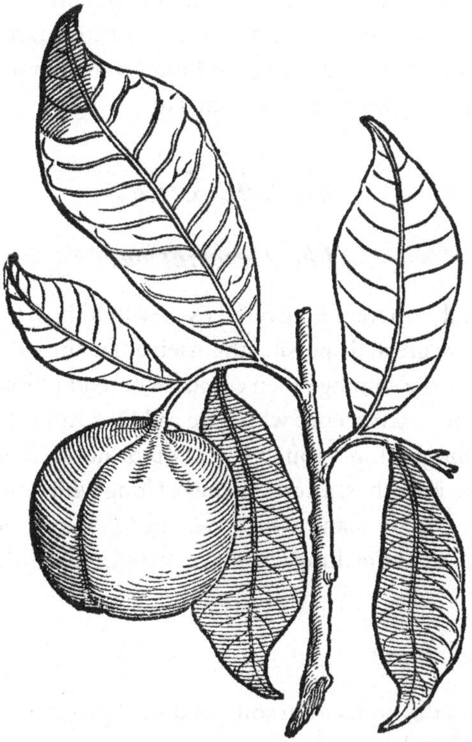

Nux myristica oblonga, siue mas.
The longiſh or male nutmeg.

nux moschatacum sua maci.

The Nutmeg with his Mace about him.

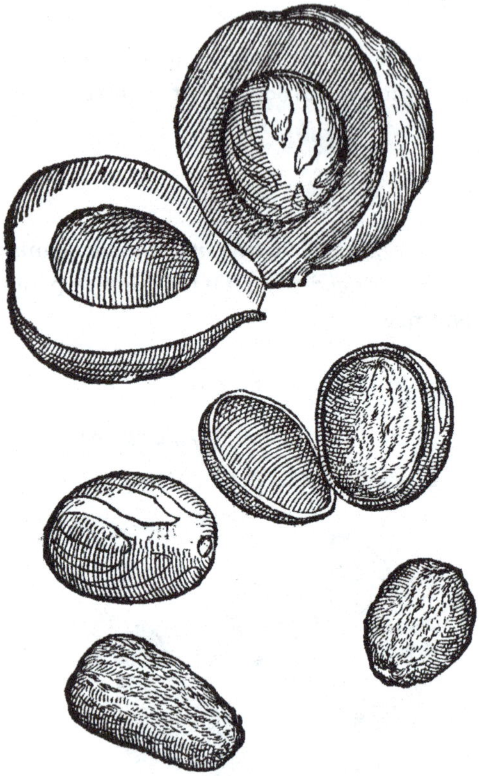

The Temperature and Vertues.

The Nutmeg, as the Mauritanians write, is hot and dry in the second degree complete, and somewhat astringent. Nutmegs cause a sweet breath and amend those that do stink, if they be much chewed and holden in the mouth. The Nutmeg is good against freckles in the face. It quickens the sight; strengthens the belly and feeble liver; takes away the swelling in the spleen; stays the laske; breaks wind; and is good against all cold diseases in the body.

Nutmegs bruised and boiled in *aqua vitæ* until they have wasted and consumed the moisture, adding thereto of *rhodomel* (that is, honey of Roses) gently boiling them, and being strained to the form of a syrup, cure all pains proceeding of windy and cold causes, if three spoonfuls be given fasting for certain days together. As touching the choice, there is not any so simple but knows that the heaviest, fattest, and fullest of juice are the best, which may easily be found out by pricking the same with a pin or such like.

Of the Oke.

The Description.

The common Oke grows to a great tree; the trunk or body whereof is covered over with a thick rough bark, full of chops or rifts. The arms or boughs are likewise great, dispersing themselves far abroad. The leaves are bluntly indented about the edges, smooth, and of a shining green colour, whereon is found a most sweet dew and somewhat clammy, and also a fungous excrescence which we call Oke Apples. The fruit is long, covered with a brown, hard and tough pilling, set in a rough scaly cup or husk. There is often found upon the body of the tree, and also upon the branches, a certain kind of long white mosse, hanging down from the same, and sometimes another woody plant, which we call Misseltoe, being either an excrescence or outgrowing from the tree it self, or of the dung (as it is reported) of a bird that has eaten a certain berrie.

The Place and Time.

The Oke does scarcely refuse any ground; for it grows in a dry and barren soil, yet does it prosper better in a fruitful ground. It grows upon hills and mountains, and likewise in valleys. It comes up every where in all parts of England, but is not so common in other of the South and hot regions. The Oke does cast his leaves, for the most part, about the end of Autumn. Some keep their leaves on, but dry all Winter long, until they be thrust off by the new Spring.

The Names.

The Oke is called in Latin, *quercus;* of some, *placida.* It may called *sativa, urbana,* or *culta;* some also name it *emeros mudion* and *robur.* We may name it in English, the tamer Oke-tree; in French, *chesne;* in Dutch, *eyeken boom.*

The fruit is named in Latin, *glans;* in High Dutch, *eichel;* in Low Dutch, *eekel;* in Spanish; *bellotus;* in Italian, *chiande;* in English, Acorne and Mast.

quercus vulgaris cum excrementus fungosis.

The common Oke with his greene Gall.

The Temperature and Vertues.

The leaves, bark, Acorn cups, and the Acorns themselves, do mightily bind and dry in the third degree, being somewhat cold withal. Acorns, if they be eaten, are hardly concocted; they yield no nourishment to man's body but that which is gross, raw, and cold. Swine are fatted herewith, and by feeding hereon, have their flesh hard and sound.

The Oke Apples are good against all fluxes of blood and laskes, in what manner soever they be taken, but the best way is to boil them in red wine, and being so prepared, they are good also against the excessive moisture and swelling of the jaws and the almonds or kernels of the throat. The decoction of Oke Apples stays women's diseases, and causes the mother that is fallen down to return again to the natural place if they do sit over the said decoction being very hot. The same steeped in strong white wine vinegar, with a little powder of Brimstone, and the root of *ireos* mingled together, and set in the Sun by the space of a month, makes the hair black, consumes proud and superfluous flesh, takes away Sun-burning, freckles, spots of the morphew, withal deformities of the face, being washed therewith.

The Oke Apples being broken in sunder about the time of their withering, do foreshow the sequel of the year, as the expert Kentish husbandmen have observed by the living things found in them; as if they find an ant, they foretell plenty of grain to ensue; if a white worm like a maggot, then they prognosticate murren of beasts and cattle; if a spider, then we shall have a pestilence or some such like sickness to follow amongst men. These things the learned also have observed and noted, that before they have a hole through them, they contain in them either a fly, a spider, or a worm. If a fly, then war ensues; if a creeping worm, then scarcity of victuals; if a running spider, then follows great sickness or mortality.

Of the great Holme-Oke.

The Description.

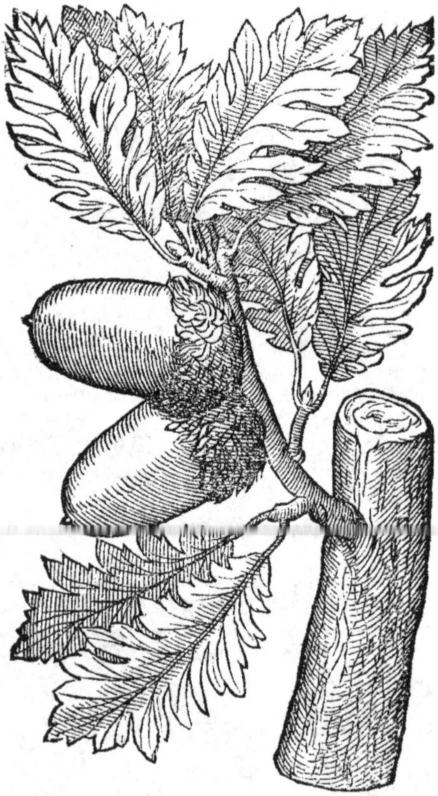

cerris maiore glande.
The Holme Oke with great Acornes.

Among the wilder Okes this is not the least for his comely proportion; although unprofitable for timber to make coles, carts, wainscot, houses, or ships of. It grows up to the height of a fair tree, the trunk or body is great and very fair to look upon. The wood or timber is soft and spongy, scarce good to be burned; from which shoot forth very comely branches dispersing themselves far abroad; whereon are set for the most part by couples very fair leaves, green above and of an overworn russet colour underneath, cut or snipt about the edges very deep.

The Acorne grows fast unto the boughs without any foot-stalk at all, being very like unto our common Acorne, set in a rough and prickly cup like a Hedge hog or the Chesnut huske, of a harsh taste, and hollow within.

This tree bears or brings forth oft times a certain smooth kind of gall not altogether unprofitable. This Oke likewise brings forth another kind of excrescence, which the Graver has omitted in the figure; Gaza names it *penis*. This penis or pricke is hollow, mossy, hanging down half a yard long, like a long rag of linen cloth. This Oke carries flowers clustering upon long stalks, like as in the common Oke; but the fruit does not succeed them, but grows forth in other places. The fruit is not fit for any man or beast to eat, neither is any property known for the use of physicke or surgery.

The Place and Time.

This Oke grows in untoiled places; it is seldom times found, and that but in woods only. It is for the most part unknown in Italy, as Pliny reports. They bring forth their fruit or Acornes in the fall of the leaf.

The Names.

This Oke is called in Latin, *cerrus*. That which hangs from the boughs, Pliny calls *panus* only.

The Temperature and Vertues.

We find nothing written of the faculties of this tree among the old writers, neither of our own experience.

Of the Scarlet Oke.

ilex coccigera.
The Scarlet Oke.

The Kindes.

Although Theophrastus has made mention but of one of these Holme or Holly Okes only, yet has the later age set down two kindes thereof, one bearing the scarlet grain, and the other only the Acorn. Which thing is not contrary to Dioscorides his opinion, for he intreats of that which bears the Acorne in his first book among the Okes; and the other he describes in his fourth book, under the title of *coccus baphice*.

The Description.

The Oke which bears the scarlet grain is a small tree, in manner of a hedge tree, of a mean bigness, having many fair branches or boughs spread abroad, whereon are set leaves, green above, white underneath, snipt about the edges, and at every corner one sharpe prickle in manner of the smoother Holly; among which comes sometimes, but not often, small Acornes standing in little cups or husks, armed with pricks as sharp as thorns, and of a bitter taste. Besides the Acornes, there is found cleaving unto the woodie branches, a certain kind of berries, or rather an excrescence of the substance of the Oke Apple, and of the bigness of a pease; at the first white, and of the colour of ashes when they be ripe, in which are ingendered little Maggots, which seem to be without life until they feel the heat of the sun, and then they creep and seek to fly away. But the people of the country (which make a gain of them) do watch the time of their flying, even as we do Bees, which they then take and put into a linen bag, wherein they shake and boult them up and down until they be dead, which they make up into great lumps often times, and likewise sell them to dyers apart, even as they were taken forth of the bag, whereof is made the most perfect Scarlet.

The Place.

This Oke grows in Languedocke, and in the countries thereabout, and also in Spain, but it bears not the scarlet grain in all places, but in those especially which lie towards the Midland sea, and which be subject to the scorching heat of the sun, as Clusius witnesses, & not there always, for when the tree waxes old, it grows to be barren. Then do the people cut and lop it down, that after the young shoots have attained two or three years growth, it may become fruitful again.

Petrus Bellonius, in his books of Singularities, shows that the scarlet grain does grow in the Holy Land and near to the lake which is called the Sea of Tiberius, and that upon little trees, whereby the inhabitants get great store of wealth, who separate the husks from the pulp or maggots, and sell this being made up into balls or lumps, much dearer than the empty shells or husks.

Theophrastus writes that the Scarlet Oke is a great tree and rises up to the height of the Common Oke, amongst which writers there are some contrarietie. Petrus Bellonius reports it is a little tree, and Theophrastus a great one, which may chance according to the soil and climate; for that upon the stony mountaines cannot grow to that greatnesse as those in the fertile grounds.

The Time.

The little grains or berries which grow about the boughs begin to appear especially in the spring, when the south west winds do blow. The floures fall and are ripe in June, together with the maggots growing in them, which receiving life by the heat of the sun, do forthwith fly away (in manner of a moth or butterfly) unless by the care and diligence of the keepers, they be killed by much and often shaking them together, as aforesaid.

The tree or shrub has his leaves always green. The Acornes be very late before they be ripe, seldom before new come up in their place.

The Names.

The Scarlet Oke is called in Latin, *ilex*. For want of a fit English name, we have thought good to call it by the name of Scarlet Oke, or Scarlet Holme Oke, for *ilex* is named of some in English, Holme, which signifies Holly. But this *ilex,* as well as those that follow, might be called Holme Oke, or Holly Oke, for difference from the shrub or hedge tree *agrifolium*.

The Arabians and the Apothecaries do know it by the name of *chesmes, chermes,* and *kermes.* The maggot within is named *cutchonele,* as most do deem.

The Temperature and Vertues.

This grain is astringent and somewhat bitter, and also dry without sharpness and biting, therefore, says Galen, it is good for great wounds and sinews that be hurt, if it be laid thereon. Some temper it with Vinegar, others with Oxymel, or syrrup of vinegar.

It is commended and given by the later physicians, to stay the menses. It is also accounted among those simples which be cordials and good to strengthen the heart. Of this graine, that noble and famous confection *alkermes,* made by the Arabians, has taken his name, which many do highly commend against the infirmities of the heart. Notwithstanding it was chiefly devised in the beginning for purging of melancholy, which thing is plainly perceived by the great quantity of *lapis lazulus* added thereto. Therefore seeing that this stone has in it a venomous quality, and likewise a property to purge melancholy, it cannot of it self be good for the heart, but the other things be good, which be therefore added that they might defend the heart from the hurts of this stone and correct the malice thereof.

This composition is commended against the trembling and shaking of the heart, and for swounings and melancholy passions, and sorrow proceeding of no evident cause. It is reported to recreate the mind, and to make a man merry and joyful.

Of the great Skarlet Oke.

The Description.

The great Skarlet Oke, or the great Holme Oke, grows many times to the full height of a tree, sometimes as big as the Peare tree, with boughs far spreading like the Acorne or common Mast trees. The timber is firm and sound. The leaves are set with prickles round about the edges, like those of the former Skarlet Oke. The leaves when the tree waxes old have on them no prickles at all, but are somewhat bluntly cut or indented about the edges, green on the upper side and gray underneath. The Acorne stands in a prickly cup like our common Oke Acorne, which when it is ripe becomes of a brown colour, with a white kernel within of taste not unpleasant. There is found upon the branches of this tree a certain kind of long, hairy mosse of the colour of ashes, not unlike to that of our English Oke.

[Thomas Johnson writes: This tree is ever green, and at the tops of the branches about the end of May, here in England, carries diverse long catkins of mossy yellow floures, which fall away and are not succeeded by the acornes, for they grow out upon other stalks. Clusius in the year 1581 observed two trees; the one in a garden above the Bridge, and the other in the private garden at White-Hall, having lesser leaves than the former. The later of these is yet standing, and every year bears small Acornes, which I could never observe to come to any maturitie.]

The Place and Time.

In diverse places there are great woods of these trees, hills also and valleys are beautified therewith. They grow plentifully in many countries of Spaine, and in Languedocke, and Provence in great plenty. It is likewise found in Italy. It bears an Acorne of a larger size than does the tame Oke; in some countries lesser and shorter. They are strangers in England, notwithstanding there is here and there a tree thereof, that has been procured from beyond the seas. One grows in her Majesties Privy Garden at White-Hall, near to the gate that leads into the street, and in some other places here and there.

It is green at all times of the year. It is late before the Acornes be ripe. Clusius reports that he saw the floures growing in clusters of a yellow colour in May.

The Names.

This Oke is named in Latin, *ilex;* in Spanish, *enzina;* in Italian, *elize;* in French, *chesneverd;* in English, Barren Skarlet Oke, or Holme Oke, and also of some, French or Spanish Oke.

The Spaniards call the fruit or Acorne, *bellota,* or *abillota.* Theophrastus seems to call this tree not *prinos,* but *smilax;* for he makes mention but of one *ilex* only, and that is of the Scarlet Oke; and he shows that the Arcadians do not call the other *ilex,* but *smilax,* for the name *smilax* is of many significations. There is *smilax* among the Pulses, and also among the Binde-weeds. Likewise *smilax* is taken of Dioscorides to be *taxus,* the Yew tree.

ilex maior glandifera.
The great Skarlet Oke.

ilicis ramus floridus.
The floures of the great Skarlet Oke.

The Temperature and Vertues.

The leaves of this Oke have force to cool and repel or keep back, as have the leaves of the Acornes or Maste trees. Being stamped or beaten, and applied, they are good for soft swellings, and to strengthen weak members.

The bark of the root boiled in water until it be dissolved, and laid on all night, makes the hair black, being first scoured with *cimolia,* as Dioscorides says.

Clusius reports that the Acorne is esteemed of, eaten, and brought into the market to be sold, in the city of Salamanca in Spaine, and in many other places of that country. Moreover, at this day in Spain, the Acorne is served for a second course.

Of Oleander, or Rose Bay.

The Description.

Rose Bay is a small shrub of a gallant show like the Bay tree, bearing leaves, thicker, greater, longer and rougher than the leaves of the Almond tree. The flowers be of a fair red colour, divided into five leaves, not much unlike a little Rose. The cod or fruit is long, and full of white down among which the seed lies hidden. The root is long, smooth, and woody.

There is another kind of Rose Bay, like the first, & differs in that this plant has white flowers; but in other respects it is very like.

nerium, siue oleander.
The Roſe Bay.

The Place.

These grow in Italy and other hot regions, by rivers and the sea side. I have them growing in my garden.

The Time.

In my garden they flower in July and August. The cods be ripe afterwards.

The Names.

This plant is called in Latin, *nerion* and also *rhododendron* and *rhododaphne;* in shops, *oleander;* in Italian, *oleandro;* in Spanish, *adelfa, eloendro,* and *alendro;* in French, *rosagine;* in English, Rose tree, Rose Bay, Rose Bay tree, and Oleander.

The Danger.

This tree, if it be inwardly taken, is deadly and poisonsome, not only to men but also to most kinds of beasts.

The flowers and leaves kill dogs, asses, mules, and very many other four-footed beasts.

The weaker sort of cattle, as sheep and goats, if they drink the water wherein the leaves have been steeped, are sure to die.

Of dwarfe Rose Bay

chamærhododendros alpigena.

Dwarfe Roſe Bay.

The Description.

Dwarfe *nerium*, or Rose Bay, has leaves which for the most part are always green, rough, and small, of a pale yellow colour like Box, far lesser than Oleander. The whole plant is of a shrubbie stature, leaning this way and that way, as not able to stand upright without help. His branches are covered and set full of small floures, of a shining scarlet or crimson colour, which being fallen, there come seed and fair berries like Asparagus.

The Place and Time.

They grow upon the hills as ye go from Trent to Verona, which in June and July are as it were, covered with a scarlet coloured carpet, of an odoriferous savour and delectable aspect.

The Names.

This may be called in English, Dwarf Rose Bay of the Alps. I find not anything extant of the vertues, so that I am constrained to leave the rest unto your own discretion.

Of the Olive tree.

The Description.

The tame or manured Olive tree grows high and great with many branches, full of long narrow leaves not much unlike the leaves of Willows, but narrower and smaller. The flowers be white and very small, growing upon clusters or bunches. The fruit is long and round, wherein is a hard stone; from which fruit is pressed that liquor which we call oile Olive.

The wild Olive is like unto the tame or garden Olive tree, saving that the leaves are something smaller; among which sometimes do grow many prickly thorns. The fruit hereof is lesser than the former, and more in number, which do seldom come to maturity or ripeness insomuch that the oile which is made of those berries continues ever green, and is called oile Omphacine, or oile of unripe Olives.

The Place and Time.

Both the tame and the wild Olive trees grow in many places of Italy, France, and Spaine, and also in the islands adjoining. They are reported to love the sea coasts, for most do think, as Columella writes, that above sixty miles from the sea they either die or else bring forth no fruit. But the best, and they that do yield the most pleasant oile, are those that grow in the island called Candy.

All the Olive trees flower in the month of June. The fruit is gathered in November or December. When they be a little dried and begin to wrinkle, they are put into the press and out of them is squeezed oile, with water added in the pressing. The Olives which are to be preserved in salt and pickle must be gathered before they be ripe and whilst they are green.

olea sativa.

The manured Olive tree.

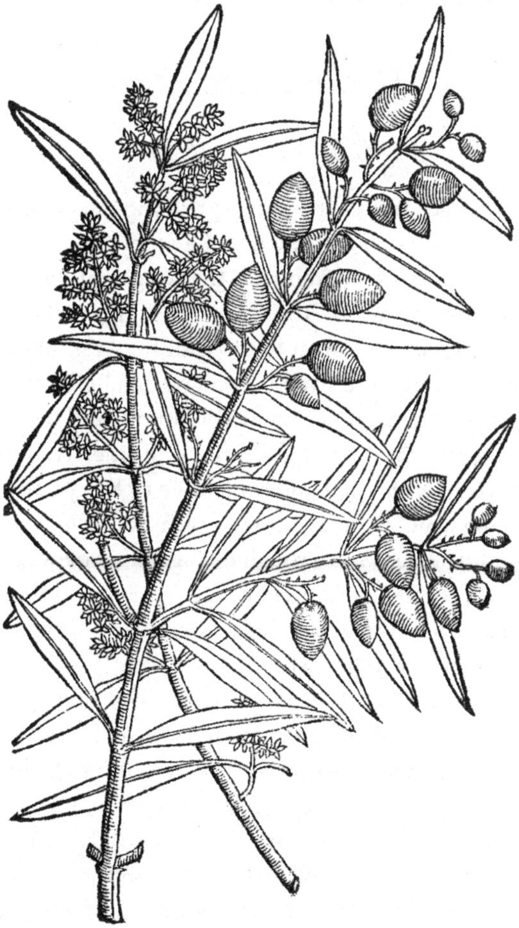

The Names.

The tame or garden Olive tree is called in Latin, *olea sativa;* in High Dutch, *delbaum;* in Italian, *olivo domestico;* in French, *olivier;* in Spanish, *olivo* and *olivera;* in English, Olive tree.

The Temperature and Vertues.

The Olives which be so ripe as that either they fall off themselves, or be ready to fall, be moderately hot and moist, yet being eaten they yield to the body little nourishment. The branches, leaves and tender buds of the Olive tree do cool, dry, and bind, and especially of the wild Olive; for they be of greater force than those of the tame. Therefore, by reason they be milder, they are better for eye medicines, which have no need of binding things to be mixed with them.

The same do stay S. Anthonies fire, the shingles, nightwheals, carbuncles, and eating ulcers. Being laid on with honey, they take away eschares, cleanse foul and filthy ulcers, quench the heat of hot swellings, and be good for kernels in the flank. Being chewed they are a remedy for ulcers in the mouth.

What is pressed forth of the unripe Olives is as cold as it is binding. The old oile, which is made of sweet and ripe Olives, being kept long, does withal become hotter and is of greater force to digest or waste away. That oile which is made of the unripe Olives, being old, does yet retain some part of his former astriction, and is of a mixed faculty, that is to say, partly binding and partly digesting. The oile of ripe Olives mollifies and assuages pain, dissolves tumours or swellings, is good for the stiffness of the joints and against cramps, especially being mingled according to art with good and wholesome herbs appropriate to those diseases and griefs, as Cammomill, Dill, Lillies, Roses, and many others, which do fortify and increase his vertues.

Of the Peach tree.

The Description.

The Peach tree is a tree of no great bigness. It sends forth diverse boughs, which be so brittle as often times they are broken with the weight of the fruit or with the wind. The leaves be long, nicked in the edges, like almost to those of the Walnut tree, and in taste bitter. The flowers be of a light purple colour. The fruit of Peaches be round, and have as it were a chink or cleft on the one side. They are covered with a soft and thin down or hairy cotton, being white without, and of a pleasant taste; in the middle whereof is a rough or rugged stone wherein is contained a kernel like unto the Almond; the meat about the stone is of a white colour. The root is tough and yellowish.

The red Peach tree is likewise a tree of no great bigness. It also sends forth diverse boughs or branches, which be very brittle. The leaves be long and nicked in the edges, like to the precedent. The flowers be also like unto the former. The fruit or Peaches be round, of a red colour on the outside; the meat likewise about the stone is of a gallant red colour. These kinds of Peaches are very like to wine in taste, and therefore marvellous pleasant.

persica alba.

The white Peach.

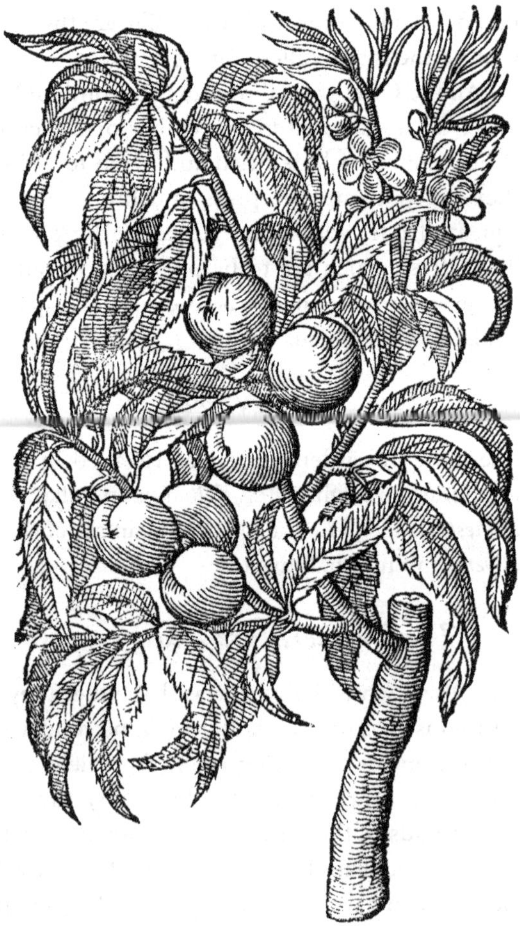

The Place and Time.

They are set and planted in gardens and vineyards. I have them all in my garden, with many other sorts. The Peach tree soon comes up. It bears fruit the third or fourth year after it is planted, and it soon decays, and is not of long continuance. It flowers in April, or a little while after that the leaves appear, and has his fruit ripe in September.

The Names.

The Peach tree is called in Latin, *malus persica;* in French, *perscher;* in English, Peach tree.

The Temperature and Vertues.

Peaches be cold and moist, and that in the second degree. They have a juice and also a substance that does easily putrefy, which yields no nourishment but brings hurt, especially if they be eaten after other meats; for then they cause the other meats to putrefy; but they are less hurtful if they are taken first, for by reason that they are moist and slippery, they easily and quickly descend, and by making the belly slippery, they cause other meats to slip down the sooner.

The kernels of Peaches be hot and dry: they open and cleanse, and are good for the stoppings of the liver and spleen.

Peaches before they are ripe do stop the laske, but being ripe they loose the belly and ingender naughty humours, for they are soon corrupted in the stomach. The leaves boiled in milk do kill the worms in children very speedily. The same being dried and cast upon green wounds, cure them. The kernel within the Peach stone stamped small and boiled with vinegar until it be brought to the form of an ointment, is good to restore and bring again the hair of such as be troubled with the *Alopecia.*

The gum is of a mean temperature, but the substance thereof is tough and clammy, by reason whereof it dulls the sharpness of thin humours. It serves in a looch or licking medicine for those that be troubled with the cough and have rotten lungs, and stops the spitting and raising up of blood, and also stays other fluxes.

Of the Peare tree.

The Description.

The Peare tree is for the most part higher than the Apple tree, having boughs not spread abroad, but growing up in height. The timber or wood it self is very tractable or easy to be wrought upon, exceeding fit to make moulds or prints to be graven on, of colour tending to yellowness. The leaf is somewhat broad, finely nicked in the edges, green above and somewhat whiter underneath. The flowers are white. The Peares, that is to say the fruit, are for the most part long, and in form like a Top; but in greatness, colour, form and taste, very much differing amongst themselves. They be also covered with skins or coats of sundry colours. The pulp or meat differs, as well in colour as taste. There is contained in them kernels, black when they be ripe. The root grows straight down with some branches running aslope.

The Place and Time.

The tame Peare trees are planted in orchards, as be the Apple trees; and by grafting, though upon wild stocks, come much variety of good and pleasant fruits. Those most rare and good are growing in the ground of Master Richard Pointer, a most cunning and curious graffer and planter of all manner of rare fruits, dwelling in a small village near London called Twicknam, and also in the ground of an excellent graffer and painful planter, Mr Henry Banbury of Touthill street near Westminster, and likewise in the ground of a diligent and most affectionate lover of plants, Mr Warner near Horsey Down, by London.

The flowers do for the most part come forth in April, the leaves afterwards. All Peares are not ripe at one time; some be ripe in July, others in August, and diverse in September and later.

The Names.

The tame or orchard Peare tree is called in Latin, *pyrus urbana;* in Low Dutch, *peerboom;* in French, *porrier.* The fruit is called in Latin, *pyrum;* in Low Dutch, *peer;* in French, *poyre;* in English, Peare.

The Temperature and Vertues.

Leaving the diverse and sundry surnames of Peares, let us come to the faculties which the Physicians ought to know, which also vary according to the differences of their tastes, for some Peares are sweet, diverse fat and unctuous, other sour, and most are harsh, especially the wilde Peares; and some consist of diverse mixtures of tastes, and some having no taste at all, but as it were a waterish taste.

All Peares are cold, and all have a binding quality and an earthy substance. All manner of Peares do bind and stop the belly, especially the harsh ones, which are good to be eaten of those that have the laske and the bloody flix. The harsh and austere Peares may with good success be laid upon hot swellings in the beginning, as may be the leaves of the tree, which do both bind and cool.

Wine made of the juice of Peares, called in English, Perry, is soluble and purges those

pirus superba, siue katherina.
The Katherine Peare tree.

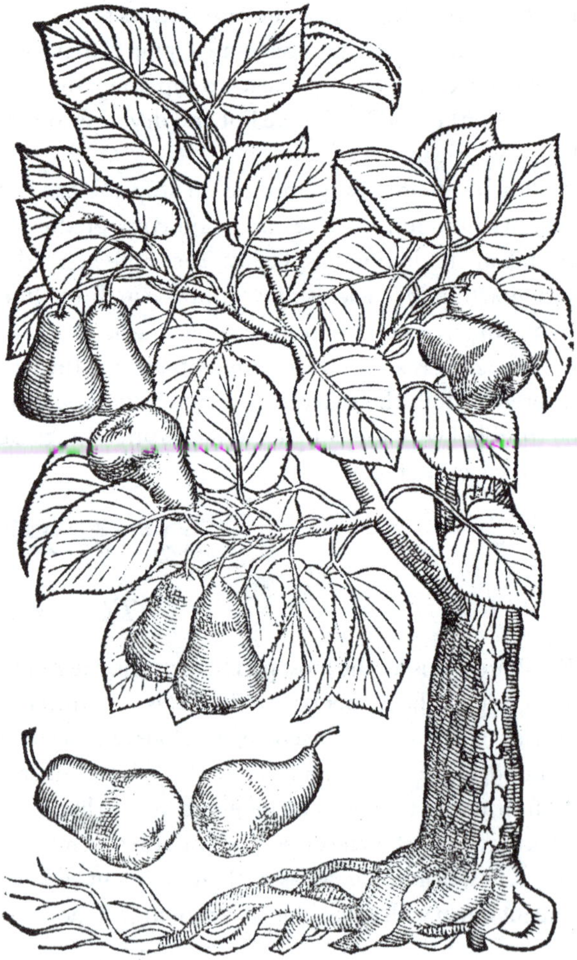

that are not accustomed to drink thereof, especially when it is new. Notwithstanding, being taken in small quantity, it is as wholesome a drink as wine; it comforts and warms the stomach and causes good digestion.

Of the wilde Peare tree.

The Kindes.

As there be sundry kindes of the manured Peare, so are there sundry wilde; whereof to write apart were to small purpose. Therefore one description with their several titles shall be sufficient for their distinctions.

The General Description.

pyrum strangulatorium maius.
The great Choke peare.

The wilde Peare tree grows likewise great, upright, full of branches, for the most part Pyramides like, or of the fashion of a steeple, not spread abroad as is the Apple or Crab tree. The timber of the trunke or body of the tree is very firme and solid, and likewise smooth, a wood very fit to make diverse sorts of instruments of, as also the hafts of sundry tools to work with withal; and likewise serves to be cut into many kinds of moulds, not only such prints as these figures are made of, but also many sorts of pretty toys, for coifes, breast-plates, and such like, used among our English gentlewomen.

The branches are smooth, covered with a blackish bark, very fragile or easy to break, whereon do grow leaves in some greater, in other lesser.

The floures are like those of the manured Pear tree, yet some whiter than others.

The fruit differ not in shape, yet some greater than others; but in taste they differ among themselves in diverse points. Some are sharpe, soure, and of an austere taste; some more pleasant, others harsh and bitter, and some of such a choking taste, that they are not to be eaten of hogs & wild beasts, much less of men. They also differ in colour, every circumstance whereof to distinguish apart would greatly enlarge our volume, and bring to the Reader small profit or commoditie.

1

2

3

4

5

1. *pyrum strangulatorium minus,* the small Choke Peare.
2. *pyrus sylvestris,* the wilde hedge Peare tree.
3. *pyrus sylvestris minima,* the wilde Crab peare tree.
4. *pyrus pedicularia,* the Lowsie wilde peare tree.
5. *pyrus corvina,* the Crow peare tree.

The Place and Time.

The wilde peares grow of themselves without manuring in most places, as woods, or in the borders of fields, and near to high ways. The time of wilde Peares answers the tame or manured Peare, notwithstanding for the most part they are not ripe much before Winter.

The Names.

The wilde Peare tree is called in Latin, *pyrus sylvestris,* by which name both the fruit and the tree are known. Peares have divers syrnames among the ancient Writers, none of which are known to the later Writers (or not desired). Every city or every country have names of themselves, and peares have also divers names according to the places.

The Temperature and Vertues.

All Peares are of a cold temperature, and the most part of them of a binding qualitie and an earthie substance. The vertues of the wilde Peares are referred to under the garden Peares as touching their binding facultie, but are not to be eaten, because their nourishment is little and bad.

Of the Pine tree.

The Description.

The Pine tree grows high and great in the trunk or body, which below is naked, but above it is clad with a multitude of boughs, which divide themselves into diverse branches, whereon are set small leaves, very straight, narrow, somewhat hard and sharp pointed. The wood or timber is hard, heavy, about the heart or middle full of an oileous liquor, & of a reddish colour. The fruit or clogs are hard, great, and consist of many sound woody scales, under which are included certain knobs, without shape, covered with a wooden shell, like small nuts, wherein are white kernels, long, very sweet, and covered with a thin skin or membrane that easily is rubbed off with the fingers; which kernel is used in medicine.

pinus sativa, siue domestica.

The tame or manured Pine tree.

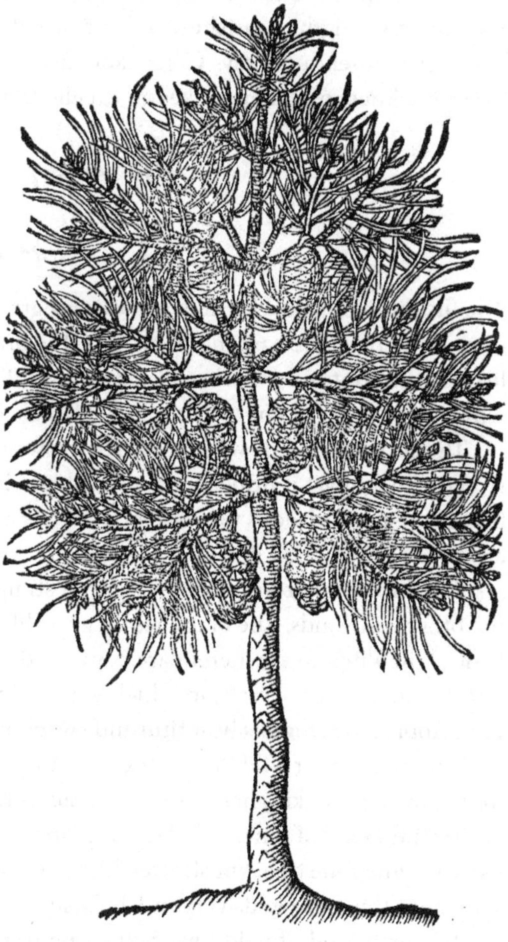

The Place.

This tree grows of it self in many places of Italy, and especially in the territory of Ravenna, and in Languedocke, about Marsiles, in Spaine, & in other regions as in the East countries. It is also cherished in the gardens of pleasure, as well in the Low-countries as England.

The Time.

The Pine tree grows green both Winter and Summer. The fruit is not commonly ripe before two years, wherefore it is not to be found without ripe fruit and also others as yet very small, and not come to ripeness.

The Names.

It is called in Latine, *pinus.* In English, tame and garden Pine. The fruits or apples of these be called in Latin, *coni;* they also be named in Latin, *nuces pineæ.* This apple is called in High Dutch, *zyzbel;* in Low Dutch, *pijn appel;* in English, Pine apple, Clogge, and Cone.

The Temperature and Vertues.

The kernels of these nuts do concoct and moderately heat, being in a mean between cold and hot. It makes the rough parts smooth; it is a remedy against an old cough and long infirmities of the chest, being taken by it self or with honey, or else with some other licking thing.

It cures the ptisicke, and those that pine and consume away though the rottenness of their lungs. It recovers strength, nourishes, and is restorative to the body. The whole Cone or Apple being boiled with fresh Horehound, says Galen, and afterward boiled with a little honey till the decoction be come to the thickness of honey, makes an excellent medicine for the cleansing of the chest and lungs.

Of the Wilde Pine tree.

The Description.

The first kind of wild Pine tree grows very great, but not so high as the former, being the tame or manured Pine tree. The bark thereof is glib. The branches are spread abroad, beset with long sharp pointed leaves. The fruit is somewhat like the tame Pine tree, with some Rosin therein, and sweet of smell, which does easily open it self and quickly falls from the tree.

The second kind of wild Pine tree grows not so high as the former, neither is the stem growing straight up, but yet it brings forth many branches, long, slender, and so easy to be bent or bowed, that hereof they make hoops for wine hogs heads and tuns. The fruit of this Pine is greater than the fruit of any of the other wilde Pines.

The third kind of wild Pine tree grows straight upright, and waxes great and high, yet not so high as the other wild kinds. The branches do grow like the Pitch tree. The fruit is long and big, almost like the fruit of the Pitch tree; wherein are contained small triangled nuts like the nuts of the Pine Apple tree, but smaller & more brittle, in which is contained a kernel of a good taste, like the kernel of the tame Pine Apple. The wood is beautiful and sweet of smell, good to make tables and other works of.

There is another wild Pine of the mountain, not differing from the precedent but in stature, growing for the most part like a hedge tree, wherein is the difference.

The fifth kind of Pine, called the sea Pine tree, grows not above the height of two men, having leaves like the tame Pine tree, but shorter. The fruit is of the same form, but longer, somewhat fashioned like a Turnep. This tree yields very much Rosin.

The sixth kind of wild Pine being one of the sea Pines, grows like a hedge tree or shrub, seldom exceeding the height of a man; with little leaves like those of the Larch tree, but always continuing with a very little cone and fine small kernel.

The bastard wild Pine tree grows up to a mean height; the trunk or body as also the branches and leaves, are like unto those of the manured Pine tree. The only difference is that some years it resembles the Pine it self, and the other years as a wild hedge tree, varying often as nature lists to play and sport her self among her delights with other plants of less moment. The timber is soft and not fit for building, but is of the substance of our Birch tree. The fruit is like those of the other wild Pines, whereof this is a kind.

The dwarf Austrian Pine exceeds not the height of a man, but immediately from the root is divided and spread abroad into tough, bending, pretty thick branches, covered over with a rough bark. The leaves as in the former, come two out of one hole, thicker, shorter, blunter pointed, and more green than the former. The cones or clogs are but small, yet round and compact, and hang not downwards but stand upright. The root is tough and woodie, like other plants of this kind. It grows on the Austrian and Styrian Alpes.

Another dwarf is of the same height as the former, with tough and bending branches, which are neither so thick nor clad with so rough a bark, nor so much spread. The leaves also are smaller and not unlike those of the Larix tree, but not so soft, nor falling every year as they do. The cones are little and slender, the kernel small, blackish, and winged as the rest. Clusius found this only in some few places, in the Kingdom of Murcia in Spaine, wherefore he calls it *pinaster.*

pinus sylvestris.
The wilde Pine tree.

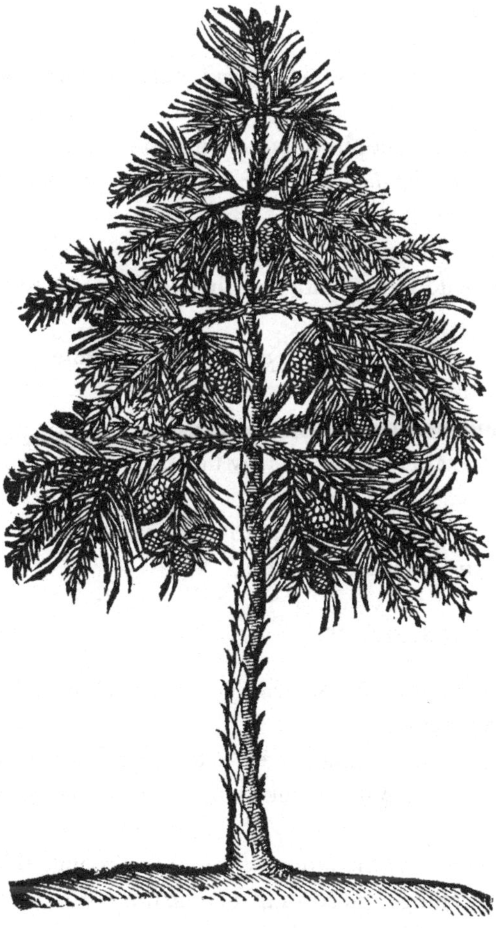

pinus sylvestris mugo.
The low wilde Pine tree.

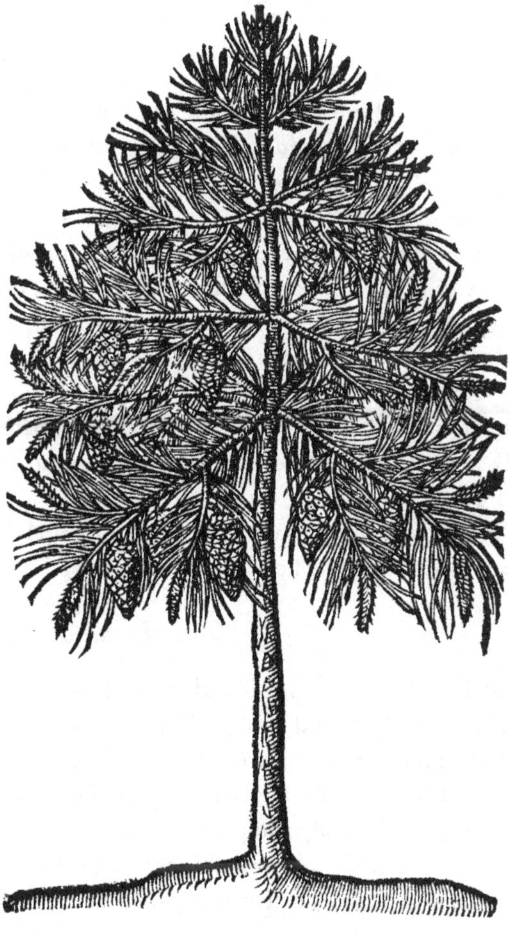

The Place.

These wild Pines do grow upon the cold mountaines of Livonia, Polonia, Norvegia, and Russia, especially upon the island called Holland within the Sownd, beyond Denmarke, and in the woods by Narua, upon the Lieseiand shore, and all the tract of the way being a thousand Wersts (each Werst containing three quarters of an English mile) from Narua unto Moscovia, where I have seen them grow in infinite numbers.

The Time.

The fruit of these Pine trees is ripe in the end of September. Out of all these issues forth a white and sweet-smelling Rosin. They are also changed into *teda,* and out of these is boiled through the force of the fire, a blacke Pitch. The Pitch tree and the Larch tree be also sometimes changed into *teda*; yet very seldom, for *teda* is a proper and peculiar infirmity of the wild Pine tree. A tree is said to be changed into *teda,* when not only the heart of it, but also the rest of the substance, is turned into fatnesse.

The Names.

All these are called in Latine, *sylvestres pini. Teda* or the Torch Pine is said to be in Latine, not *picea,* but *pinea,* that is, not the Pitch tree but the Pine tree.

The first of these wild Pines may be *idæa theophrasti,* or Theophrastus his Pine tree, growing on Mount Ida. This wild Pine tree is called in High Dutch, *hartzbaum* and *wilder hartzbaum;* in Spanish, *pino carax.*

The second wild Pine tree is named commonly of the Italians, *tridentinis* and *cembro* and *cirmolo.* It seems to differ nothing at all from the Macedonians' wild female Pine, for the wood is easy to be wrought on and serves for divers and sundry works.

The third they call *mugo.* This may be named not without cause *humilis pinus* or dwarfe Pine; yet it differs from *Chamæpeuce,* the herbe called in English, Ground Pine.

The fourth wild Pine is named in Latine, *maritima* and *marina pinus.* In English, Sea Pine.

That which the Latines call *teda* is named in High Dutch, *kynholtz.* It may be termed in English, Torch Pine.

Pliny is deceived in that he supposes the Torch Pine to be a tree by it self, and makes it the sixth kind of cone tree as likewise he errs in taking *larix,* the Larch tree, for the Pine tree.

The Temperatures and Vertues.

The bark of them both does bind; being beaten and applied it cures merigals, and also shallow ulcers and burnings if it be laid on with Litharge and fine Frankincense.

With the cerote of Myrtles it heals ulcers in tender bodies. Being beaten with Copperas it stays tetters and creeping ulcers. It draws away the birth and after birth if it be taken under in a fume. Being drunk, it stops the belly and provokes urine.

Galen has almost the same things, but he says that the bark of the Pine tree is more temperate than that of the Pitch tree; the leaves stamped take away hot swellings and sores that come thereof.

Being stamped and boiled in vinegar they assuage the pain of the teeth, if they be washed with this decoction hot. The same be also good for those that have bad livers, being drunk with water or mead. Of the same operation is likewise the bark of the Pine nuts; but Galen affirms that the cone or apple, although it seems to be like these, is notwithstanding of lesser force, insomuch as it cannot effectually perform any of the aforesaid vertues, but has in it a certain biting quality, which hurts.

The Torch Pine cut into small pieces and boiled in vinegar, is a remedy likewise against the tooth-ache, if the teeth be washed with the decoction. Of this there is made a profitable spather or slice, to be used in making of compound plaisters and pessaries that ease pain. Of the smoke of this, is made a black which serves to make ink of, and for eating sores in the corner of eyes, and against the falling away of the eye lids, and for watering and bleere eyes, as Dioscorides teaches.

Of the Pitch tree.

The Description.

Picea, the tree that drops the Pitch, called Pitch tree, grows up to be a tall, fair, and big tree, remaining always green like the Pine tree. The timber of it is more red than that of the Pine or Fir. It is set full of boughs, not only about the top, but much lower, and also beneath the middle part of the body, which many times hang down, bending toward the ground. The leaves be narrow, not like those of the Pine tree, but shorter and narrower, and sharp pointed like them, yet are they blacker, and withal cover the young and tender twigs in manner of a circle, like those of the Fir tree; but being many and thick set, grow forth on all sides, and not only one right against anther, as in the Yew tree. The fruit is scaly, and like unto the Pine apple, but smaller. The bark of the tree is somewhat black, tough and flexible, not brittle, as is the bark of the Fir tree; under which, next to the wood is gathered a Rosin, which many times issues forth, and is like to that of the Larch tree.

The Place and Time.

The Pitch tree grows in Greece, Italy, France, Germany and all the cold regions even unto Russia. The fruit of the Pitch tree is ripe in the end of September.

picea maior.
The Pitch tree.

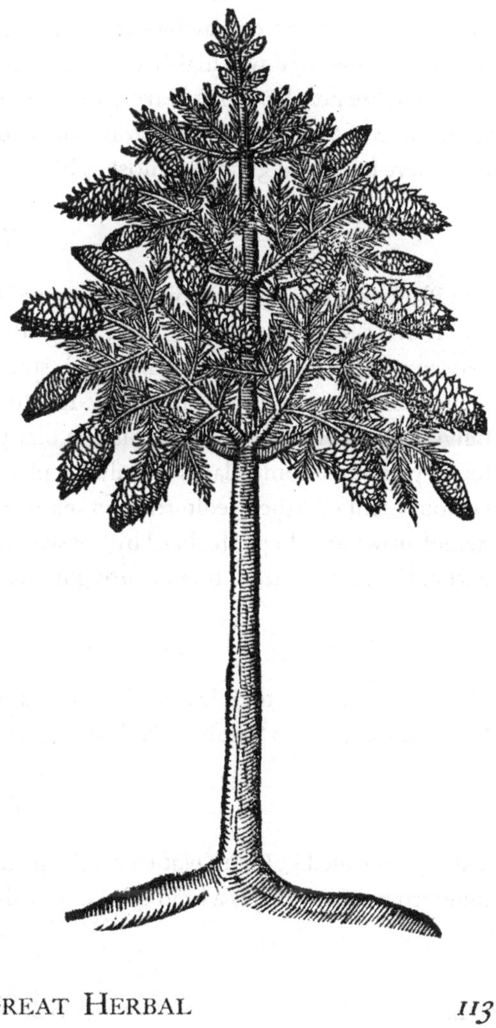

The Names.

The Latins call this tree, *picea.* It is named in High Dutch, *schwartz tannenbaum,* and *rot tannenbaum,* and oftentimes also *jozenholtz;* which name notwithstanding does also agree with other plants. In Low Dutch, *peck boom;* in English, Pitch tree.

The Temperature and Vertues.

The leaves, bark, and fruit of the Pitch tree are all of one nature, vertue and operation, and of the same faculty with the Pine trees.

Of the Plane tree.

The Description.

The Plane is a great tree, having very long and far spreading boughs casting a wonderful broad shadow, by reason whereof it was highly commended and esteemed of among the old Romans. The leaves are cornered, greater than Vine leaves, and hanging upon little red foot-stalks. The flowers are small and mossy, and of a pale yellowish colour. The fruit is round like a ball, rugged, and somewhat hairy; but in Asia more hairy and greater, almost as big as a Walnut. The root is great, dispersing it self far abroad.

The Place.

The Plane tree delights to grow by springs or rivers. Pliny reports that they grew to be of so great honour as that they were cherished and watered with wine; and it is found by experience that the same is very comfortable to the roots. This tree is strange in Italy, it is nowhere seen in Germany, nor in the Low countries. In Asia it grows plentifully. It is found also in Candy, growing in valleys, and near unto the hill Athos, as Petrus Bellonius in his Singularities does declare. It grows in many places of Greece, and is found planted in some places of Italy, for pleasure rather than for profit. My servant, William Marshall, (whom I sent into the Mediterranean sea as surgeon unto the Hercules of London) found diverse trees hereof growing in Lepanto, hard by the sea side, at the entrance into the town, a port of Morea, being a part of Greece, and from thence brought one of those rough buttons, being the fruit thereof.

The Time.

The Plane trees cast their leaves in Winter, as Bellonius testifies, and therefore it is no marvel that they keep away the sun in Summer, and not at all in Winter.

The Names.

This tree is called in Latin, *platanus;* it bears his name of the breadth. The French men's *plasne* does far differ from this, which is a kind of Maple. This tree is named in English, Plane tree.

platanus .

The Plane tree .

The Temperature and Vertues.

The Plane tree is of a cold and moist essence, as Galen says. The green leaves are good to be laid upon hot swellings and inflammations in the beginning. Being boiled in wine, they are a remedy for the running and the watering of the eyes, if they be applied. The bark and balls do dry. The bark boiled in vinegar helps the tooth ache. The fruit of the Plane tree drunk with wine helps the bitings of mad dogs and serpents, and mixed with hogs' grease, it makes a good ointment against burning and scalding. The burned bark does mightily dry and scours withal; it removes the white scurf and cures moist ulcers.

Of the Plum tree.

The Description.

prunus domestica.
The Damſon tree.

The Plum or Damson tree is of a mean bigness. It is covered with a smooth bark. The branches are long, whereon do grow broad leaves, more long than round, nicked in the edges. The flowers are white. The Plums consist of pulp and skin, and also of kernel, which is shut up in a shell or stone. Some Plums are of blackish blew, of which some be longer, others rounder, others of the colour of yellow wax, diverse of a crimson red, greater for the most part than the rest. There be also green Plums, and withal very long, of a sweet and pleasant taste. Moreover, the pulp or meat of some is drier, and easilier separated from the stone. Our common Damson is known to all, and therefore not to be stood upon.

The Place and Time.

The Plum trees grow in all known countries of the world. They require a loose ground, they also receive a difference from the regions where they grow, not only of the form or fashion, but especially of the faculties. The Plum trees are also many times graffed into trees of other kinds, and being so ingraffed they *faciem parentis, succum adoptionis, ut Plinius dicit, exhibent.* The greatest variety of these rare Plums are to be found in the grounds of Mr Vincent Pointer of Twicknam, although myself am not without some, and those rare and delicate. The wild Plums grow in most hedges through England.

The common and garden Plum trees do bloom in April. The leaves come forth presently with them. The fruit is ripe in Summer, some sooner, some later.

The Names.

The Plum tree is called in Latin, *prunus;* in Spanish, *ciruelo;* in French, *prunier;* in English, Plum tree. The fruit is called in Latin, *prunum;* in Italian and French, *prune;* in Spanish, *prunas;* in English, Prune and Plum.

The Temperature and Vertues.

Plums that be ripe and new gathered from the tree, what sort soever they are of, do moisten and cool, and yield unto the body very little nourishment and the same nothing good at all. For as Plums do very quickly rot, so is also the juice of them apt to putrefy in the body, and likewise cause the meat to putrefy which is taken with them. Only they are good for those that would keep their bodies soluble and cool; for by their moisture and slipperiness they do mollify the belly.

Dried plums, commonly called prunes, are wholesomer and more pleasant to the stomach; they yield more nourishment and better, and such as cannot easily putrefy. Galen affirms in his books of the Faculties of Simple Medicines, that they do manifestly loose the belly. Being boiled with mead or honeyed water, which has a good quantity of honey in it, they loose the belly very much, although a man take them alone by themselves, and much more if the mead be supped after them.

The leaves of the Plum tree are good against the swelling of the uvula, the throat, gums, & kernels under the throat and jaws. They stop the rheum and falling down of humours, if the decoction thereof be made in wine, and gargled in the mouth and throat.

Of Sebesten, or the Assyrian Plum.

sebestenae, myxa, siue myxara .

Aſſyrian Plums.

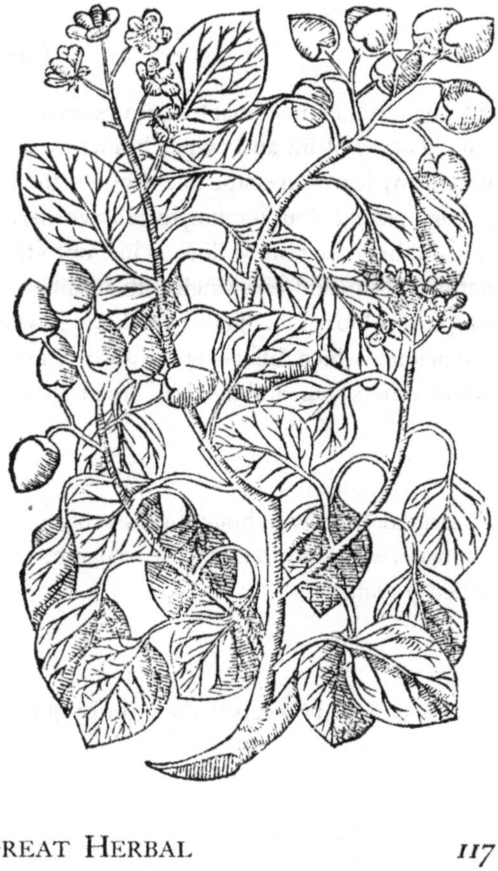

The Description.

The Sebestines are also a kind of Plums, the tree whereof is not unlike the Plum tree, saving it grows lower than most of the manured Plum trees and the leaves be harder and rounder. The flowers grow at the tops of the branches consisting of five small white leaves, with pale yellow threads in the middle, like those of the Plum tree. After follows the fruit like to little plums, fastened in little skinny cups, which when they be ripe are of a greenish black colour, wherein is contained a small hard stone. The fruit is sweet in taste, the pulp or meat is very tough and clammie.

The Place and Time.

The Sebesten trees grow plentifully in Syria and Egypt; they were in times past forreine and strange in Italy, now they grow almost in every garden, being first brought thither in Pliny's time. Now do the Sebesten trees, says he, begin to grow in Rome, among the Service trees. The time answers the common Plums.

The Names.

Pliny called the tree *myxa.* The berry or fruit is named *myxon* and *myxarion,* neither have the Latines any other name. The Arabians and the Apothecaries do call it *sebesten,* which is also made an English name. We may call it the Assyrian Plum.

The Temperature and Vertues.

Sebestens be very temperately cold and moist, and have a thick and clammie substance; therefore they nourish more than most fruits do, but withal they easily stop the intrailes, and stuff up the narrow passages, and breed inflammations.

They take away the ruggedness of the throat and lungs, and also quench thirst, being taken in a looch or licking medicine, or prepared any other kind of way, or else taken by themselves. The weight of ten drams, or of an ounce and a halfe of the pap or pulpe hereof being inwardly taken, does loose the belly.

There is also made of this fruit a purging electuarie, but such a one as quickly moulds and therefore it is not to be used but when it is new made.

Of Pomegranat tree.

The Description.

The manured Pomegranat tree grows up to the height of a hedge tree, being seven or eight cubits high, having many pliant and twiggy branches, very limber, tough, and of a brown colour, whereon are set very many leaves in shape like those of the Privet, but more like those of the Myrtle tree, of a bright shining green colour tending to yellowness; among which there stand certain sharp thorns confusedly set, and likewise hollow flowers like unto the hedge Rose, indented on the edges like a star, of a carnation colour, and very single, after which comes the fruit, covered with a hard bark, of an overworn purplish colour, full of grains and kernels, which after they be ripe are of a gallant crimson colour, and full of juice, which differs in taste according to the soil, climate, and country where they grow. Some be sweet, others sour, and the third are in the middle between them, having the taste of wine.

The Place.

Pomegranats grow in hot countries toward the South, in Italy, Spaine, and chiefly in the kingdom of Granado, which is thought to be so named of the great multitude of Pomegranats, which be commonly called *granata.*

The Time.

The Pomegranat flowers in the months of May and June. The fruit is ripe in the end of August.

malus granata, siue punica.
The Pomegranat tree.

The Names.

The Pomegranat tree is called in Latin, *Malus Punica;* in English, Pomegranate tree. The fruit is named in Latin, *malum punicum;* in shops, *malum* or *pomum granatum;* in Italian, *melagrano* and *pomo granato;* in Spanish, *granadas;* in French, *pommes granades;* in English, Pomegranate.

The Temperature and Vertues.

The juicy grains of the Pomegranate are good to be eaten, having in them a meetly good juice. They are wholesome for the stomach, but they all contain in them a thin and small nourishment, or none at all.

The sweet ones be not so cold as the rest, but they easily cause hot swellings to arise, and they are not so much commended for agues.

The sour ones, and especially if they be withal something harsh, do evidently cool, dry, and something bind.

They are good for the heart burn. They repress and stay the overmuch vomiting of choler, called the felonie. They are a remedy against the bloody flix, aptness to vomit, and vomit it self.

There is made of the juice of these sour Pomegranates a syrup, which serves for the same purposes, and is also many times very profitable against the longing of women with child, unless the coldness of the stomach be a hindrance thereto.

The seeds or grains, and especially of the sour Pomegranate, being dried, do likewise cool and bind. They stop the flix, stay vomiting, and stanch the spitting up of blood. They strengthen the stomach.

Of the same effect be the flowers, both of the tame and wild Pomegranate tree, being like to the seeds in temperature and vertues. They fasten the teeth and strengthen the gums, if the same be washed therewith. They are good against the burstings that come by falling down of the guts, if they be used in plaisters and applied.

The seeds or stones of Pomegranates dried in the sun and beaten to powder, are of like operation to the flowers. They stop the laske and all issues of blood in man or woman, being taken in the manner aforesaid.

Of the Poplar tree.

The Description.

The white Poplar tree comes soon to perfection, and grows high in short time, full of boughs at the top. The bark of the body is smooth, and that of the boughs is like wise white withal. The wood is white, easy to be cleft. The leaves are broad, deeply gathered, and cornered like almost to those of the Vine, but much lesser, smooth on the upper side, glib, and somewhat green; and on the nether side white and woolly. The catkins are long, downy, at the first of a purplish colour. The roots spread many ways, lying under the turf and not growing deep, and therefore it happens that these trees be often times blown down with the wind.

populus alba.

The white Poplar tree.

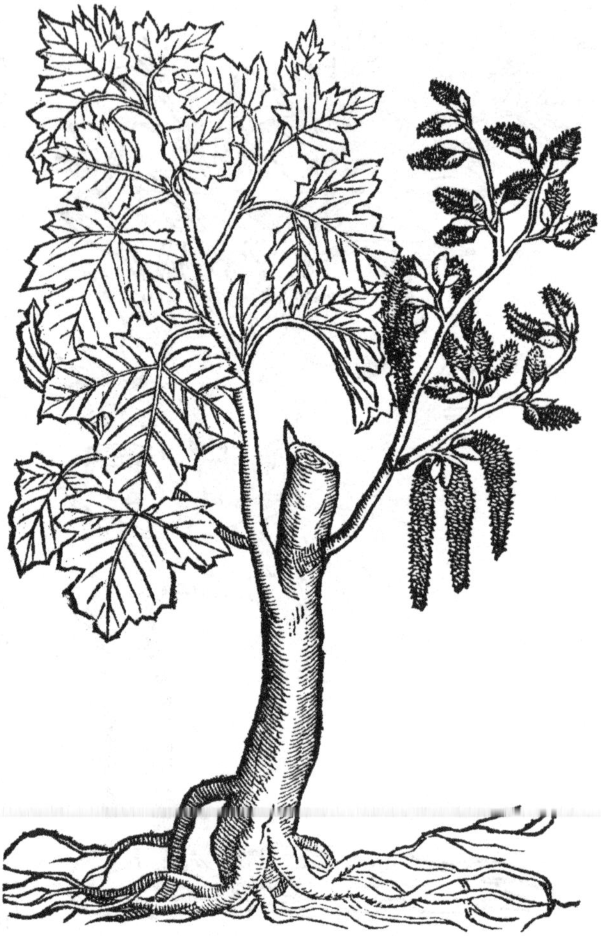

The Place and Time.

The white Poplar grows not very common in England, but in some places here and there a tree. I found many both small and great growing in a low meadow turning up a lane at the farther end of a village called Black-wall, from London, and in Essex at a place called Ovenden, and in diverse other places.

These trees do bud forth in the end of March and beginning of April, at which time the buds must be gathered to serve for *unguentum populeon.*

The Names.

The white Poplar is called in Latin, *populus alba.* It is called in High Dutch, *poppelbaum;* in Low Dutch, *abeel,* of his hoary or aged colour, and also *abeelboome;* which the Grammerians do falsely interpret *abies,* the Firre Tree. It is called in Italian, *popolo nero;* in French, *peuplier blanc;* in English, white Poplar tree and Abeell after the Dutch name.

The Temperature and Vertues.

The white Poplar has a cleansing faculty, says Galen, and a mixed temperature, consisting of a watery warm essence, and also of a thin earthy substance. The bark, as Disocorides writes, to the weight of an ounce (or as others say, and that more truly, of little more than a dram) is a good remedy for the Sciatica or ache in the huckle bones, and for the strangerie. That this bark is good for the Sciatica, Serenus Sammonicus does also write:

A hidden disease does oft rage and rain,
The hip overcome and vex'd with the pain,
It makes with vile aching one tread slow, and shrink;
The bark of white Poplar is help had in drink.

The same bark is also reported to make a woman barren, if it be drunk with the kidney of a mule, which thing the leaves likewise are thought to perform, being taken after the flowers or reds be ended. The warm juice of the leaves being dropped into the ears does take away the pain thereof.

Of the Prickly Indian Fig tree.

The Description.

This strange and admirable plant, called *ficus indica,* seems to be no other thing than a multiplication of leaves, that is, a tree made of leaves, without body or boughs; for the leaf set in the ground does in short space take root and brings out of it self other leaves, from which do grow others one after another, till such time as they come to the height of a tree, having also in the mean season boughs, as it were, coming from those leaves, sometimes more, otherwhiles fewer, as nature list to bestow, adding leaf unto leaf, whereby it occupies a great piece of ground.

These leaves are long and broad, as thick as a man's thumb, of a deep green colour, set full of long slender sharp and whitish prickles. On the tops of which leaves come forth long flowers, not unlike to those of the manured Pomegranate tree, of a yellow colour. After which comes the fruit like unto the common Fig, narrow below and bigger above, of a green colour, and stuffed full of a red pulp and juice, staining the hands of them that touch it, as do the Mulberries, with a bloody or sanguine colour; the top of which Figs are invironed with certain scaly leaves like a crown, wherein are also contained small grains that are the seeds; the which being sown do bring forth plants round bodied, like unto the trunk of other trees, with leaves placed thereon like the other; which being set in the ground bring forth trees of leaves, as we have shown.

Upon this plant in some parts of the West Indies grow certain excrescences, which in continuance of time turn into insects; and these out-growings are that high prized Cochenele wherewith they dye colours in grain.

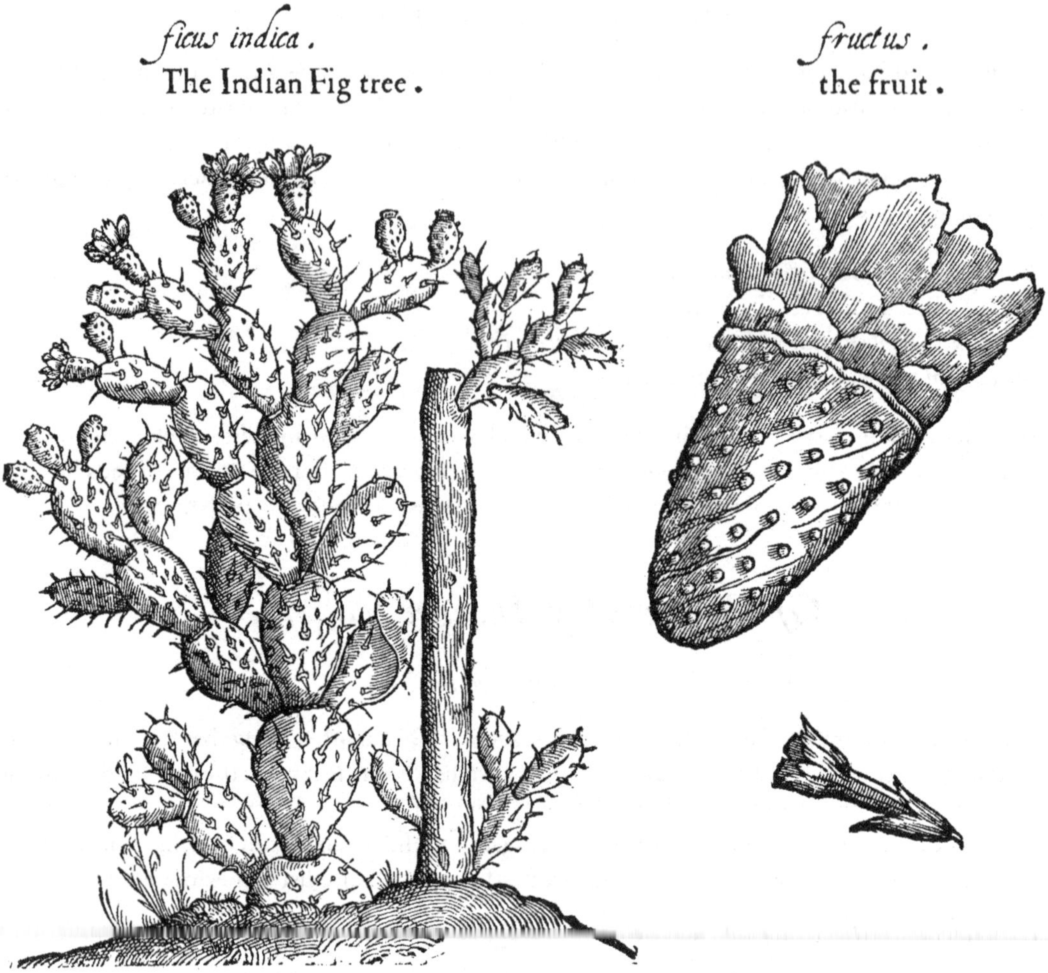

The Place and Time.

This plant grows in all the tract of the East and West Indies, and also in the country Norembega, now called Virginia, from whence it has been brought into Italy, Spaine, England and other countries. In Italy, it sometimes bears fruit, but more often in Spaine and never as yet in England, although I have bestowed great pains and cost in keeping it from the injury of our cold climate.

It grows also at S. Crux and other places of Barbary, and also in an island of the Mediterranean sea called Zante, about a day and night's sailing with a mean wind from Petrasse, a port in Morea where my servant William Marshall did see not only great store of those trees made of leaves, but also diverse other round bodied plants of a woody substance; from whence he brought me diverse plants thereof in tubs of earth, very fresh and green, which flourish in my garden at the impression hereof. These plants do grow green and fresh both Winter and Summer, by the relation of my foresaid servant. Notwithstanding, they must be very carefully kept in these countries from the extremity of Winter.

The Names.

It is commonly called in Latin, *ficus indica;* of the Indians, *tune* and *tunas* and also *anapallus,* as Bellonius testifies. In English, Indian Fig tree.

The Temperature and Vertues.

We have no certain instruction from the ancients of the temperature or faculty of this plant, or of the fruit thereof, neither have we any thing whereof to write of our own knowledge, more than that we have heard reported of such as have eaten liberally of the fruit hereof, that it changed their urine to the colour of blood; who at the first sight thereof stood in great doubt of their life thinking it had been blood, whereas it proved afterward by experience to be nothing but the tincture or colour the urine had taken from the juice of the fruit, and that without all hurt or grief at all.

Of Privet, or Prim Print.

The Description.

Privet is a shrub growing like a hedge tree, the branches and twigs whereof be straight, and covered with soft glistering leaves of a deep green colour like those of Peruincle, but yet longer, greater also than the leaves of the Olive tree. The flowers be white, sweet of smell, very little, growing in clusters; which being faded there succeed clusters of berries, at the first green and when they be ripe, blacke like a little cluster of grapes, which yield a purple juice. The root grows every way aslope.

ligustrum.
Privet, or Prim Print.

The Place.

The common Privet grows naturally in every wood, and in the hedge rows of our London gardens. It is not found in the country of Polonia and other parts adjacent.

The Time.

It flowers in the end of May, or in June; the berries are ripe in Autumne or about Winter, which now and then continue all the Winter long; but in the meane time the leaves fall away, and in the Spring new come up in their places.

The Names.

It is called in Latine, *ligustrum.* Serapio the Arabian does call Privet, *mahaleb.* There is also another *mahaleb* which is a graine or seed of which Avicen makes mention that it does by his warme and comfortable heate dissolve and assuage paine. Serapio seems to intreat of them both, and to containe diverse of the *mahaleb* under the title of one chapter. It is named in High Dutch, *beinholtzlein, mundtholtz, rhein oder schulweiden* and in French, *troesne.* In English, Privet, Primprint and Print.

Some there be that would have the berries to be called *vaccinia,* and *vaccinium* to be that of which Vitruvius has made mention in his seventh book of Architecture or the art of building, of purple colours; after the same manner, says he, they temper *vaccinium* and putting milk unto it do make a gallant purple. In such brevitie of the old writers what can be certainly determined.

The Temperature and Vertues.

The leaves and fruit of Privet are cold, dry, and astringent. The leaves do cure the swellings, apostumations, and ulcers of the mouth or throat, being gargarised with the juice or decoction thereof, and therefore they be excellent good to be put into lotions to wash the secret parts, and the scaldings with women, and cankers and sores in children's mouths.

Of the Quince tree.

The Description.

The Quince tree is not great, but grows low, and many times in manner of a shrub. It is covered with a rugged bark, which has on it now and then certain scales. It spreads his boughs in compass like other trees, about which stand leaves somewhat round, like those of the common Apple tree, green and smooth above, and underneath soft and white. The flowers be of a white purple colour, the fruit is like an Apple, save that many times it has certain embowed and swelling divisions. It differs in fashion and bigness, for some Quinces are lesser and round trussed up together at the top with wrinkles, others longer and greater.

The Place and Time.

The Quince tree grows in gardens and orchards, and is planted often times in hedges and fences belonging to gardens and vineyards. It delights to grow in plain and even grounds, and somewhat moist withal. These Apples be ripe in the fall of the leaf, and chiefly in October.

The Names.

The tree is called in Latin, *malus cotonea;* in English, Quince tree. The fruit is named *malum cotoneum, pomum cydoneum,* and many times, *cydonium* without any addition, by which name it is made known to the Apothecaries. It is called in High Dutch, *kuttenopffel;* in Italian, *mele cotogne;* in Spanish, *codoyons;* in French, *pomme de coing;* in English, Quince.

malus cotonea.
The Quince tree.

Quinces be cold and dry in the second degree, and also very much binding, especially when they be raw. They have likewise in them a certain superfluous and excremental moisture, which will not suffer them to lie long without rotting. They are seldom eaten raw; being roasted or baked, they be more pleasant.

Simeon Sethi writes that the woman with child, which eats many Quinces during the time of her breeding, shall bring forth wise children, and of good understanding.

The Marmalade, or Cotiniate, made of Quinces and sugar, is good and profitable for the strengthening of the stomach, that it may retain and keep the meat therein until it be perfectly digested. It likewise stays all kinds of fluxes, both of the belly and other parts, and also of blood; which Cotiniate is made in this manner:

Take fair Quinces, pare them, cut them in pieces and cast away the core, then put into every pound of Quinces a pound of sugar; and to every pound of sugar, a pint of water. These must be boiled together over a still fire till they be very soft, then let it be strained or rather rubbed through a strainer, or a hairy sieve, which is better, and then set it over the fire to boil again until it be stiff, and so box it up, and as it cools put thereto a little Rose water and a few grains of Muske, well-mingled together, which will give a goodly taste unto the Cotiniate.

Many other excellent, dainty and wholesome confections are to be made of Quinces, as jelly of Quinces, and such odd conceits, which for brevity sake I do now let pass.

Of Rosins.

The Kindes.

Out of the Pine trees, especially of the wild kinds, there issues forth a liquid, whitish, and sweet smelling Rosin, and that many times by it self; but more plentifully out of the cut or broken boughs, or forth of the body when the tree comes to be a torch Pine. There issues also forth of the cracks and chinks of the bark, or out of the cut boughs, a certain dry Rosin, and that forth of the Pine tree or Fir tree. There is likewise found a certain congealed Rosin upon the cones or apples.

It is called in Latin, *resina;* in High Dutch, *hartz;* in Low Dutch, *herst;* in Italian, *ragia;* in Spanish, *resina;* and in English, Rosin.

The Temperature and Vertues.

All the Rosins are hot and dry, but not all after one manner, for there is a difference among them. They which be sharper and more biting, are hotter, as that which comes of the cones, being of Rosins the hottest, because it is also the sharpest. The Rosin of the Pitch tree is not so much biting, and therefore not so hot. The Rosin of the Fir tree is in a mean between them both; the liquid Rosin of the Pine is moister, coming near to the quality of the Larch Rosin.

The Rosins which are burned or dried, as Dioscorides testifies, are profitable in plaisters, and compositions that ease wearisomeness; for they do not only supple or mollify, but also by reason of the thinness of their parts and dryness, they digest. Therefore they both mollify and waste away swellings, and through the same faculty they cure wearisomeness, being used in compound medicine for that purpose.

The liquid Rosins are very fitly mixed in ointments, commended for the healing up of green wounds, for they both bring to suppuration, and do also glue and unite them together.

Of Pitch and Tar.

The Manner and Drawing Forth of Pitch.

Out of the fattest wood of the Pine tree, changed into the torch Pine, is drawn Pitch by force of fire. A place must be paved with stone or some other hard matter, a little higher in the middle, about which there must also be made gutters, into which the liquor must fall; then out from them other gutters are to be drawn, by which it may be received. Being received, it is put into barrels. The place being thus prepared, the cloven wood of the torch Pine must be set upright; then must it be covered with a great number of Fir and Pitch boughs, and on every part all about with much loam and earth; and great heed must be taken, lest there be any cleft or chink remaining, only a hole left in the top of the furnace, through which the fire may be put in, and the flame and smoke may pass out. When the fire burns, the Pitch runs forth, first the thin, and then the thicker.

This liquor is called in Latin, *pix.* In English, Pitch, and the moisture, even the same that first runs, is named of Pliny, *cedria.* There is boiled in Europe, says he, from the torch Pine, a liquid pitch used about ships, and serving for many other purposes; the wood being cloven is burned with fire, and set round about the furnaces on every side, after the manner of making charcoals. The first liquor runs through the gutter like water (this is called in Syria, *cedrium,* which is of so great vertue, as in Ægypt the bodies of dead men are preserved, being all covered over with it).

The liquor following, being now thicker, is made Pitch; and of this, when it is boiled again, is made a harder pitch named *iterata pix* or Pitch iterated, because it is boiled a second time. A certain kind hereof being clammy or gluing is named in shops, *pix navalis;* in English, Ship Pitch or Stone Pitch.

The Temperature and Vertues.

Pitch is hot and dry, tar is hotter, and Stone Pitch more drying. Tar is good against inflammations of the almonds of the throat and uvula, and likewise the Squincie, being outwardly applied. It is a remedy for mattering ears with oile of Roses, and heals the bitings of serpents if it be beaten with salt and applied. With an equal portion of wax, it takes away foul, ill favoured nails; it wastes away swellings of the kernels and hard swellings of the mother and fundament.

Of the Sassafras or Ague tree.

The Description.

The Sassafras tree grows very great, much like to the pine tree. The trunk or body is great, smooth, and void of boughs, of a great height. It is covered with a twofold gross rind, the uppermost of the colour of ashes, that next the wood of a tawny colour.

On the top come forth many goodly branches like those of the Palm tree, whereon grow green leaves somewhat like those of the Fig tree, of a sweet smell when they be green, but much sweeter when they be dry, declining to the smell of Fennel, with much sweetness in taste. They are green Winter and Summer, neither bearing fruit nor flowers, but is altogether barren as it is said.

The roots are grosse, conformable to the greatness of the tree, of a tawny colour, dispersing themselves far abroad under the upper crust of the earth, by means whereof they are often cast down with mean blasts of wind.

The wood of the tree is very strong, hard and brittle, it has not so strong & pleasant a smell as that

sussafras .

The Saſſafras tree.

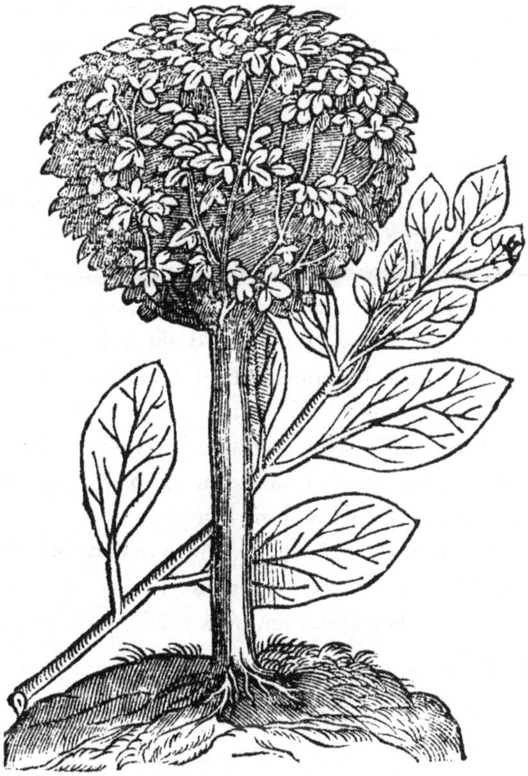

of the root, neither is it in such use. The leaves are of two sorts, some long and smooth, and not snipt about the edges; othersome, and those chiefly on the end of the branches, are deeply gashed in, as it were divided into three several parts. I have been given the figure of a branch taken from a little tree, which grew in the garden of Mr Wilmot at Bow, who died some few years ago.

The Place.

This tree grows in most parts of the West Indies, especially about the cape of Florida, Wingandico, and Virginia, otherwise named Norembega.

The Time.

It flourishes and keeps green Winter and Summer.

The Names.

The Spanish and French men have named this tree *sassafras.* The Indians in their tongue, *pauame.* For want of an English name we are contented to call it the Ague tree, of his vertue in healing the Ague.

The Temperature.

The boughs and branches hereof are hot & dry in the second degree. The rind is hotter for that it enters into the third degree of heat and dryness, as is manifestly received in the decoction.

The Vertues.

The best of all the tree is the root, and that which works the best effect has the rind cleaving very fast to the inner part and is of colour tawny, and much more sweet of smell than all the tree and his branches.

It is a tree that grows near unto the sea, and in temperate places that have not much drought, nor moisture. There be mountaines growing full of them, and they cast forth a most sweet smell, so that at the beginning when men saw them first, they thought they had been trees of Cinnamon, & in part they were not deceived. For the rind of this tree has as sweet a smell as Cinnamon has, and does imitate it in colour and sharpness of taste & pleasantnesse of smell, and procures the same works and effects as Cinnamon does.

The wood hereof cut in small pieces and boiled in water, to the colour of Claret wine, and drunk for certain days together, helps the dropsie, removes oppilation or stopping of the liver, cures quotidian and tertian agues, and long fevers.

The root of Sassafras has power to comfort the liver and the weak and feeble stomach, to cause good appetite, to consume windinesse, to stay vomiting and make sweet a stinking breath.

It provokes urine, removes the impediments that do cause barrenness, and makes women apt to conceive.

Of Savin.

The Description.

There be two kinds of Savin; one like in leaf to Tamariske, the other to the Cypresse tree, whereof the one bears berries, the other is barren.

The first Savin, which is the common kind and best known in this country, grows in manner of a low shrub or tree, the stem or trunk whereof is sometimes as big as a man's arm, dividing it selfe into many small branches set full of small leaves like unto Cypress or Tamariske, but thicker and more sharpe or prickley, remaining green Winter and Summer, in smell rank or very strong, barren both of floures and fruit.

The other Savin is a high tree, as tall as the Almond tree, and much like to the tame Cypress tree. The body is writhed, thick and sometimes of so great a compass as that it cannot be fathomed. The substance of the wood is red within, as is that of the Juniper, and of the prickely Cedar. The bark is not very

Sabina sterilis.
Barren Savin.

Sabina baccifera.
Savin bearing berries.

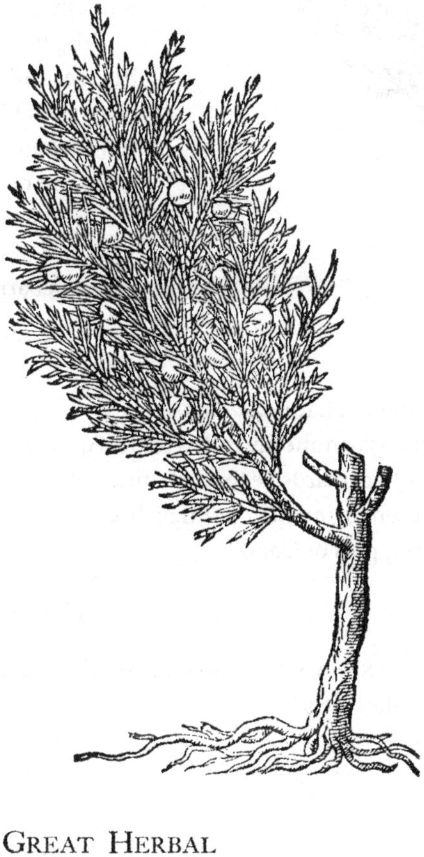

sabina baccata altera.

The leſſer berry-bearing Savin.

thick and is of a yellowish red. The leaves are of a marvelous gallant green colour, like to those of the Cypress tree, yet thicker or more in number; in taste bitter, of a spicie smell, and like Rosin.

The boughes are broader, and thick set as it were with wings, like those of the Pitch tree and of the Yew tree, on which grow a great number of berries, very round like those of the little Cedars, which at the first are green but when they be ripe they are of a blackish blew. Out of the root issues oftentimes a rosin, which being hard is like to that of the Juniper tree, and does also crumble in the chewing.

There is another which differs from the last described only in that the leaves are smaller and less pricking, as also the branches lesser. Lobel calls this *savina baccata altera.*

The Place.

Both of them grow upon hills, in woods, and in other untoiled places, as in Candy, Mysia and elsewhere. P. Bellonius reports that he found them both upon the tops of the mountaines Taurus, Amanus and Olympus.

The first is planted in our English gardens almost every where. The second is planted both by the seed and by the slip. The slips must be set in a ground that is meanly moist and shadowie till they have taken root. The shrubs which grow out of these decline toward the one side, retaining still the nature of the bough, but that Savin which is planted by the seed grows more upright. This in continuance of time brings forth seeds, and the other for the most part remains barren. Both these grow in my garden.

The Names.

Savin is called in Latine, *sabina.*

The first is commonly called in the Apothecaries shops by the name *savina.* Of divers, *savimera.* The Italians and Spaniards keep the Latine name. It is called in high-Dutch, *sibenbaum;* in low-Dutch, *savelboom;* in French, *savenier;* in English, common Savine or garden Savine. Some name the other *cupressus cretica,* or Cypres of Candy.

The Temperature.

The leaves of Savin, which are most used in medicine, are hot and dry in the third degree, and of subtill parts, as Galen says.

The Vertues.

The leaves of Savin boiled in wine and drunk provoke urine, bring down the menses with force, draw away the afterbirth, expel the dead child, and kill the quicke. It has the like vertue received under in a perfume.

The leaves stamped with honey and applied cure ulcers, stay spreading and creeping ulcers, scoure and take away all spots and freckles from the face or body of man or woman. The leaves boiled in oyle Olive and kept therein kill the worms in children if you anoint their bellies therewith, and the leaves powdered and given in milk or Muscadell do the same.

The leaves dried and beat into fine powder and strewed upon those kinds of excrescences *sub præputio* and such like, gotten by dealing with uncleane women, take them away perfectly, curing and healing them. But if they be inveterate and old and have been much tampered withal, it shall be necessary to add unto the same a small quantitie of *Auripigmentum* in fine powder, and use it with discretion, because the force of the medicine is greatly increased thereby, and made more corrosive.

Of the Service tree.

The Description.

The Service tree grows to the height and bignesse of a great tree, charged with many great arms or boughs which are set with sundry small branches, garnished with many great leaves somewhat long like those of the Ash. The floures are white and stand in clusters, which turn into small brown berries somewhat long, which are not good to be eaten until they have lain a while and until they be soft like the Medlar, whereto it is like in taste and operation.

The common Service tree grows likewise to the height of a great tree, with a straight body of a brownish colour, full of branches, set with large displayed leaves like the Maple or the White-Thorne, saving that they are broader and longer. The floures are white and grow in tufts, which being fallen there come in place thereof small round berries, browne upon one side and red toward the Sun, of an unpleasant taste in respect of the former; in which are contained little blackish kernels.

The Place and Time.

These trees are found in woods and groves in most places of England. There be many small trees thereof in a little wood a mile beyond Islington from London. In Kent it grows in great aboundance, especially about Southfleet and Gravesend.

They floure in March and their fruit is ripe in September.

The Names.

The first is called in Latine, *sorbus.* In high-Dutch, *sperwerbaum;* in low-Dutch, *sozbedboom;* in French, *cormier;* in English, Service tree and of some after the Latine, Sorbe tree. The common Service tree is named of Pliny, *sorbus torminalis.*

sorbus.

The Service tree.

sorbus torminalis.

Common Service tree.

The Temperature and Vertues.

Service berries are cold and binding, and much more when they be hard then when they are mild and soft. In some places they are quickly soft, either hanged in a place which is not altogether cold, or laid in hay or chaffe. Those Services are eaten when the belly is too soluble, for they stay the same, and if they yield any nourishment at all, the same is very little, gross and cold, and therefore it is not expedient to eat of these or other-like fruits, nor to use them otherwise than in medicines.

These do stay all manners of fluxes of the belly, and likewise the bloody flixes, as also vomiting. They stanch bleeding if they be cut and dried in the sun before they be ripe, and so reserved for use. These we may use divers ways according to the manner of the grief and grieved part.

Of the Spindle tree or Pricke-wood

The Description.

The English Prick-timber tree, or Prickewood, is no high shrub, of the bigness of the Pomegranat tree. It spreads far with his branches. The old stalks have their bark somewhat white, the new and those that be lately grown be green and four square. The substance of the wood is hard and mixed with a light yellow. The leaves be long, broad, slender and soft. The floures be white, many standing upon one foot-stalk, like almost to a spoked rundle. The fruit is four square, red, and containing four white seeds, every one whereof is covered with a yellow coat which being taken off gives a yellow dye.

The Broad leafed Spindle tree is another sort of Euonymous which grows to the form of a hedge tree, of a mean bigness, the trunk or body whereof is of the thickness of a man's leg, covered with a rough or scabbed bark of an overworn russet colour. The branches thereof are many, slender and very even, covered with a green bark whilst they be yet young and tender. They are also very brittle, with

euonymous theophrasti.
Englifh Prick-timber tree.

euonymous latifolius.
Broad leafed Spindle tree.

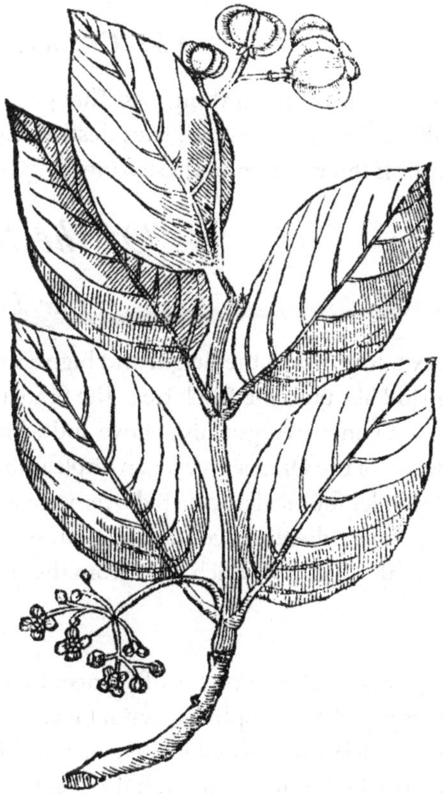

some pith in the middle like that of the Elder. The leaves are few in number, full of nerves or sinews dispersed like those of Plantain, in shape like those of the Pomecitron tree, of a loathsome smell and a bitter taste. Amongst the leaves come forth slender footstalks, very long and naked, whereon do grow small floures consisting of four small leaves like those of the Cherrie tree, but lesser, of a white colour tending to a blush with some yellowness in the middle. After comes the fruit, which is larger than the former, and as it were winged, parted commonly into four, yet sometimes into five parts. Opening when it is ripe, it shows the white grains filled with a yellow pulp. The root is tough and woody, dispersing it self far abroad under the upper crust of the earth.

The Place and Time.

It comes up in untoiled places and among shrubs, upon rough banks and heaps of earth. It serves also oftentimes for hedges in fields, growing amongst brambles and such other thorns.

The floures appear in April. The fruit is ripe in the end of August or in the month of September.

The Names.

It is called in High Dutch, *spindelbaum;* in low Dutch, *papenhout;* in Italian, *fusano;* in French, *fusin* and *bonnet depestre;* in English, Spindle tree, Prick wood and Prick-timber.

The Temperature and Vertues.

This shrub is hurtful to all things, as Theophprastus writes, and namely to Goats; he says the fruit hereof kills, as do the leaves and fruit destroy Goats especially, unless they scoure as well upwards as downwards. If three or four of these fruits be given to a man, they purge both by vomit and stool.

Of the Storax tree.

The Description.

The Storax tree grows to the height and bigness of the Quince tree. The trunk or body is covered with a bark or rind like unto the Birch tree. The branches are small and limmer, whereon do grow leaves like those of the Quince tree, greenish above and whitish underneath; among which come forth white flowers, like those of the Orange tree, of an unpleasant smell. After comes the fruit or berries, standing upon long and slender foot-stalks, covered over with a little woolliness, of the bigness of a bladder nut and of the same colour; wherein is contained small seed, whereunto also cleave certain gummy tears, bearing the name of the tree, and which issue from the trunk or body when it is wounded.

The Place.

This tree grows in diverse places of France, Italy and Spaine, where it brings forth little or no gum at all. It grows in Judæa, Pamphylia, Syria, Pisidia, Sidon, and many other places of Jewrie or Palestine, as also in diverse islands in the Mediterranean sea, namely Cyprus, Candy, Zant, and other places, where it brings forth his gummy liquor in full perfection of sweetness and also in great plenty, where it is gathered and put into great Canes or Reeds, whereof as some deem it took the name *calamita;* others deem of the leaves of Reeds wherein they wrap it. Hereof I have two small trees in my garden, the which I raised of seed.

styrax arbor.

The Storax tree.

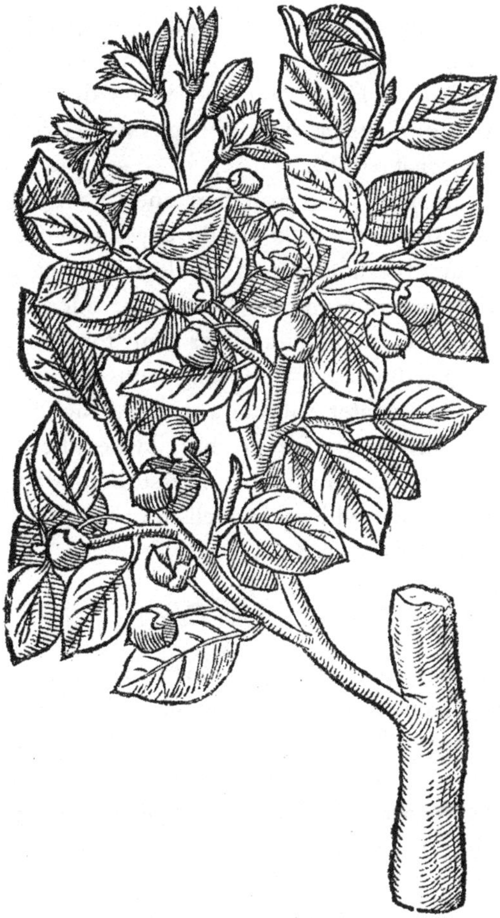

The Time.

It flowers in May, and the fruit is ripe in September.

The Names.

This tree, as may be gathered by some, was called Styrax, by reason of that gum or liquor which drops out of the same, being like unto the hollow pipes of ice that hang at the eaves of houses in Winter, called *styria,* or of the Canes or the leaves of Reeds spoken of before. In Latin, *storax calamitæ;* in English, Storax. There flows from some of these trees a certain gummy liquor, which never grows naturally hard but remains always thin, which is called liquid Styrax or Storax.

The Temperature and Vertues.

The gum of this tree is of a heating, mollifying, and concocting quality. It helps the cough, the falling down of rheums and humours into the chest, and hoarseness of the voice. It also helps the noise and sounding of the ears, prevails against *strumas,* or the King's Evill, nodes on the nerves, and hard swellings proceeding of a cold cause. It prevails also against all cold poisons, as Hemlocks and such like.

Of this gum there are made sundry excellent perfumes, pomanders, sweet waters, sweet bags, and sweet washing balls, and diverse other sweet chains and bracelets, whereof to write were impertinent to this history.

Of the Strawberry Tree.

arbutus.

The Strawberry tree.

The Description.

The Strawberry tree grows for the most part low, very like in bigness to the Quince tree (whereunto Dioscorides compares it.) The body is covered with a reddish barke, both rough and scaly. The boughs stand thick on the top, somewhat reddish.

The leaves be broad, long and smooth like those of Bays, somewhat nicked in the edges, and of a pale green colour. The floures grow in clusters being hollow and white, and now and then on the one side somewhat of a purple colour. In their places come forth certain berries hanging down upon little long stems like unto Strawberries, but greater, without a stone within, but only with little seeds, at the first green, and when they be ripe they are of a gallant red colour, in taste somewhat harsh and in a manner without any relish; of which Thrushes and Black-birds do feed in Winter.

The Place.

The Strawberry tree grows in most countries of Greece, in Candy, Italy and Spaine, also in the valleys of the mountain Athos, where being in other places but little, they become great huge trees as P. Bellonius writes. Iuba also reports that there be in Arabia of them fifty cubits high. They grow only in some few gardens with us.

The Time.

The Strawberry tree floures in July and August and the fruit is ripe in September, after it has remained on the tree by the space of a whole year.

The Names.

This tree is called in Latine, *arbutus;* in English, Strawberry tree and of some, Arbute tree. The fruit is named in Latine, *memæcylum* and *arbutus* and Pliny called it *unido.* The Italians call this Strawberry *albatro;* the Spaniards, *madrono;* the French, *arboutes.* It may be termed in English, Tree Strawberry.

The Temperature and Vertues.

The fruit of the Strawberry tree is of a cold temperature, hurting the stomacke and causing head-ache. Wherefore no wholesome food, though it be eaten in some places by the poorer sort of people.

Of Coriars Sumach.

The Description.

Coriars Sumach grows up to the height of a hedge tree, after the manner of the Elder tree, bigger than Dioscorides reports it to be, or others who affirm that *rhus* grows two cubits high, whose errors are the greater. But this *rhus* is so like to the Service tree in shape and manner of growing that it is hard to know one from the other; but that the leaves are soft and hairy, having a red sinew or rib through the midst of the leaf. The floures grow with the leaves upon long stems clustering together like cats taile, or the catkins of the nut tree, but greater and of a whitish green colour; after which come clusters of round berries growing in bunches like grapes.

Sumach of Pliny's description grows like a small hedge tree, having many slender twiggie branches, garnished with little leaves like Myrtus, or rather like the leaves of the Iuibe tree, among which come forth slender mossie floures of no great account or value, which bring forth small seeds, enclosed within a cornered case or husk, fashioned like a spoon.

The trunk or body of both these kinds of Sumach being wounded with some iron Instrument, yields a gum or liquor.

rhus coriaria.
Coriar Sumach.

The Place.

Sumach grows, as Dioscorides says, in stony places. It is found in diverse mountaines & woods in Spaine, and in many places on the mount Apennine in Italy, and also near unto Pontus. Archigenes in Galen, in the 8th book of medicines according to the places affected, shows that it grows in Syria.

The Time.

The floures of Sumach come forth in July; the seed with the berries are ripe in Autumn.

The Names.

Rhus, says Pliny, has no Latine name. The Arabians name it *sumach;* the Italians *sumacho;* the Spaniards, *sumagre;* in Low Dutch, *smack* or *sumach.* The leaves of the shrub be called in Latine, *rhus coriaria* or *rhoe.*

The seed is named *eruthros;* in Latine, *rhus culinaria* and *rhus obsoniorum;* in English, Meat Sumach and Sauce Sumach.

The Temperatures.

The fruit, leaves and seeds hereof do very much binde; they also cool and dry. Dry they are in the third degree and cold in the second, as Galen teaches.

The Vertues.

The leaves of Sumach boiled in wine and drunken, do stop the laske, the inordinate course of women's sicknesses and all other inordinate issues of blood.

The seed of Sumach eaten in sauces with meat stops all manner of fluxes of the belly, the bloody flix, and all other issues, especially the white issues of women.

The decoction of the leaves makes hairs black and is put into stooles to fume upward into the bodies of those that have the Dysenterie, and is to be given them also to drink.

The leaves made into an ointment or plaister with honey and vinegar, stays the spreading nature of of Gangrenes and Pterygia.

The dry leaves sodden in water until the decoction be as thick as honey, yields forth a certain oilinesse which performs all the effects of Licium.

The seeds pounded, mixed with honey and the powder of Oaken coals heals the Hemorrhoides.

There issues out of the shrub a gum, which being put into the hollowness of the teeth, takes away the pain, as Dioscorides writes.

Of red Sumach.

The Description.

These two figures are of one and the selfsame plant: the first shows the shrub being in flower; the other when it is full flowered with the fruit grown to ripeness, notwithstanding some have deemed them to be of two kindes, wherein they were deceived.

This excellent and most beautiful plant *coggygria* (being reputed of the Italians and the Venetians for a kind of *rhus* or Sumach, because it is used for the same purposes whereto *rhus* serves and therein does far excel it) is an hedge plant growing not above the height of four or five cubits, having tough and pliant stalks and twiggie branches like to Oziers, of a brown colour. The leaves be brown, thick and stiff like the leaves of *capparis,* in colour and savour of *pistacia* leaves or *terebinthus,* among which arises a small upright sprig bearing many small clustering little greenish yellow flowers, upon long and red

coggygria theophrasti.
Venice Sumach.

cotinus coriarius pliny.
Red Sumach.

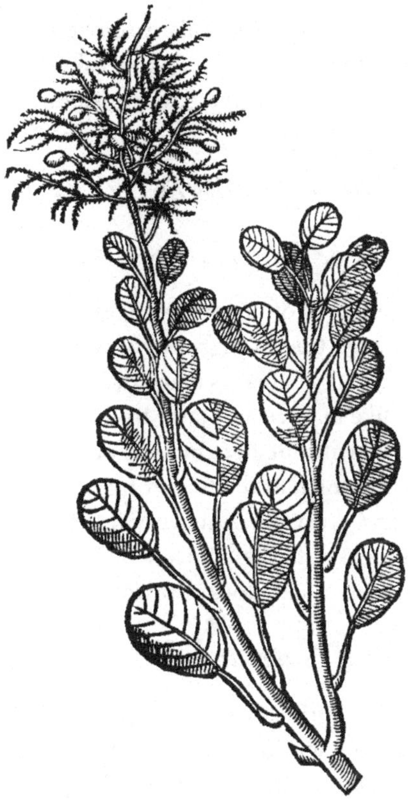

stalks. After which follow small, reddish Lentil-like seeds that carry at the tops a most fine woolly or flockie tuft, crisped and curled like a curious wrought silken fleece, which curls and folds it self abroad like a large bush of hairs.

The Place and Time.

Coggygria grows in Orleans near Avignion and in diverse places of Italy, upon the Alpes of Styria, and many other places. It grows on most of the hills of France, in the high woods of the upper Pannonia or Austria, and also of Hungaria and Bohemia. They floure and flourish for the most part in July.

The Names.

The first is called *coggygria* and *coccygria;* in English it is called Venice Sumach or Silken Sumach. Divers would have named it *scotinus,* which name is not found in any of the old writers. The Pannonians call it *farblauff.*

The Temperatures and Vertues.

The leaves and slender branches together with the seeds are very much binding, cold and dry as the other kinds of Sumach are.

The leaves of *coggygria,* or Silken Sumach, are sold in the markets of Spaine and Italy for great sums of money, unto those that dress Spanish skins, for which purpose they are very excellent. The roots of *cotinus* serves to dye with, giving to wool and cloth a reddish colour.

Of the Sycomore tree.

sycomorus.
The Sycomore tree.

The Description.

The Sycomore tree is of no small height being very like the Mulberrie tree in bignesse & shew, as also in leafe. The fruit is as great as a fig, and of the same fashion, very like in juice and taste to the wilde Fig, but sweeter and without any grains or seeds within; which grows not forth of the tender boughs but out of the body and great old arms very fruitfully. This tree has in it plenty of milkie juice which so soon as any part is broken or cut, does issue forth.

The Place and Time.

It grows, as Dioscorides writes, very plentifully in Caria and Rhodes and in sundry places of Egypt as at the great Cayre or Alkaire, and in places that do not bring forth much wheat, in which it is a help and suffices in stead of bread & corne when there is scarcitie of victuals.

It brings forth fruit three or four times in one year, and oftener if it be scraped with an iron knife or other like instrument.

The Names.

This tree is called in Latine, *sycomorus.* The Egyptians of our time do call it *ficus pharaonis,* or Pharao his Fig tree, as witnesses Bellonius. It is likewise termed *ficus ægyptia,* Egyptian Fig tree, and also *morus ægyptia,* or Egyptian Mulberrie tree. We call it in English, Sycomore tree after the Greek and Latine, and also Mulberry Fig tree, which is the right Sycomore tree and not the great Maple as we have said in the chapter of the Maple. The fruit is named in Greek, *sycomoron,* and in Italian, *sycomoro* and *fico d'egitto.*

The Temperature and Vertues.

The fruit of the Sycomore tree has no sharpness in it at all, as Galen says. It is somewhat sweet in taste and is of temperature moist after a sort, and cold as be Mulberries.

It is good, says Dioscorides, for the belly; but it is without any nourishment and troublesome to the stomacke.

There issues forth of the bark of this tree in the beginning of the Spring, before the fruit appears, a liquor, which being taken up with a sponge or a little wool, is dried, made up into fine cakes, and kept in gallie pots. This mollifies, closes wounds together, and dissolves grosse humours.

It is both inwardly taken and outwardly applied against the bitings of serpents, hardness of the milt or spleen, and pain of the stomacke proceeding of a cold cause. This liquor does very quickly putrifie.

Of Tamariske.

The Description.

The first kind of Tamariske grows like a small hedge tree, covered with a reddish bark, having many branches set and bedecked with leaves, much like unto Heath. Among which come forth small mossie white flowers declining to purple, which turn into a pappous or downy seed that flies away with the wind as that of Willow does. The root is woody as the roots of other shrubs be, and grows diverse ways.

The Germane Tamariske has many woody branches or shoots rising from the root, with a white bark, having his leaves thicker and grosser than the former, and not so finely jagged or cut. The flowers are reddish and larger than the former, growing each a pretty distance from another on the tops of the branches, spike fashion. They begin to flower below, which do turn into seed that is likewise carried away by the wind.

The Place and Time.

Tamariske grows by running streams and many times by rivers that break forth, and not seldom about fenny grounds, commonly in a gravelly soil, for it best prospers in moist and stony places. It is found in Germany, Vindelicia, Italy, Spaine, and also in Greece.

The Tamariskes do also grow in Egypt and Syria, as Dioscorides writes; and likewise in Tylus, an island in Arabia, as Theophrastus notes. The wood whereof, says he, is not weak as with us, but strong, like timber or any other strong thing. Dioscorides does call this Tamariske tame or planted, and says that it brings forth fruit very like to Galls, in taste rough and binding.

Petrus Bellonius, in his second book of singularities, reports that he saw in Egypt very high Tamariskes, and that some times in moist places by rivers sides, and many times also in dry and gravelly grounds where no other trees did grow, which now and then did bear hanging on the boughs such a multitude of Galls, as being over loden they were ready to break. Both these grow and prosper well in gardens with us here in England.

These trees or shrubs floure in May, and in the latter end of August their seed is carried away with the wind.

tamariscus narbonensis.
French Tamariske.

tamariscus germanica.
Germane Tamariske.

The Names.

They are called in Latine, *myrica* and *tamarix;* in shops, *tamariscus.* It is named in High Dutch, *tamarischen holk* and *pozk;* in Italian, *tamarigio;* in Spanish, *tamarguira* and *tamariz;* in French, *tamaris;* in English, Tamariske.

The Temperature and Vertues.

Tamariske has a cleansing and cutting facultie with a manifest drying. It is also somewhat astringent or binding and by reason of these qualities it is very good for an hard spleen, being boyled with vinegar or wine, either the root or leaves or tender branches, as Galen writes.

Moreover, Dioscorides teaches that the decoction of the leaves made with wine does waste the spleen and that the same is good against the tooth ache if the mouth be washed therewith. It brings down the Menses if the patient sit therein and it kills lice and nits if the parts be bathed therewith.

The ashes of burnt Tamariske has a drying facultie, and greatly scouring withal, and a little binding.

The floures and downie seed of the greater Tamariske does greatly bind, insomuch as it comes very near to the Gall named *galla omphacitis*, but that the roughness of taste is more evident in the Gall. The floures are of an unequal temperature, for there is joined to the nature thereof a great thinnesse of parts and cleansing facultie which the Gall has not, as Galen writes.

These floures we fitly use, in stead of Gall, in medicines for the eyes and mouth. It is good to stanch blood and to stay the laske and women's whites, it helps the yellow jaundice and also cures those that are bit of the venomous spider called *phalangium*. The bark serves for the same purposes.

The leaves and wood of the Tamariske have great power and virtue against the hardness and stopping of the spleen, especially the leaves being boiled in water and the decoction drunk, or else infused in a small vessel of Ale or Beere and continually drunk; and if it be drunk forth of a cup or dish made of the wood or timber of the Tamariske, is of greater efficacie.

Of the Tree of Life.

The Description.

The Tree of Life grows to the height of a small tree, the bark being of a dark reddish colour, the timber very hard, the branches spreading themselves abroad, hanging down toward the ground by reason of the weakness of the twiggie branches surcharged with very oileous and ponderous leaves, casting and spreading themselves like the feathers of a wing, resembling those of the Savine tree, but thicker, broader and more full of gummie or oileous substance. Which leaves, being rubbed in the hands, do yield an aromatic, spicie or gummie savour, very pleasant and comfortable.

Amongst the leaves come forth small yellow floures, which in my garden fall away without any fruit, but as it has been reported by those that have seen the same, there follows a fruit in hot regions, much like unto the fruit of the Cypresse tree, but smaller, compact of little and thin scales closely pact one upon the other, which my selfe have not yet seen.

The branches of this tree laid down in the earth will very easily take root, even like the Woodbinde or some such plant, which I have often proved and thereby have greatly multiplied these trees.

arbor vitæ.
The Tree of Life.

The Time and Place.

This tree groweth not wild in England, but it grows in my garden very plentifully.

It endures the cold of our Northern climate, yet does it lose his gallant greens in the winter months. It floures in my garden about May.

The Names.

Theophrastus and Pliny, as some think, have called this sweet and aromatical tree *thuia,* or *thya.* Some call it *cedrus lycia.* The new writers do term it *arbor vitæ;* in English, the Tree of Life. I do not mean that whereof mention is made in Genesis 3:22.

The Temperature and Vertues.

Both the leaves and boughs be hot and dry.

Among the plants of the New-found land, this tree is the most principal and best agreeing unto the nature of man, as an excellent cordial and of a very pleasant smell.

Of the Turpentine tree.

The Description.

The first Turpentine Tree grows to the height of a tall and faire tree, having many long boughes or branches dispersed abroad, beset with long leaves, consisting of sundry other small leaves, each whereof resembles the Bay leafe, growing one against another upon a little stem or middle rib, like unto the leaves of the Ash tree.

The floures be small & reddish, growing upon clusters or bunches that turn into round berries which at their beginning are green, afterwards reddish, but being ripe wax blacke or of a dark blew colour, clammie, full of fat, oileous in substance and of a pleasant savour.

This plant bears an empty cod, or crooked horn, somewhat reddish, wherein are found small flies, worms or gnats, bred and ingendered of a certain humorous matter which cleaves to the inner sides of the said cods or horns, which worms have no physical use at all.

The right Turpentine issues out of the branches of these trees if you do cut or wound them; the which is faire and clear and better than that which is gathered from the bark of the Firre tree.

The second kind of Turpentine tree is very like unto the former but that it grows not so great, yet the leaves are greater and broader and of the same fashion, but very like to the leaves of the Pistacia tree. The berries are first of a scarlet colour, and when they be ripe of a skie colour.

The great horned cods are sharp pointed and somewhat cornered, consisting as it were of the substance of gristles. And out of those bladders being broken do creep and come small flies or gnats, bred of a fuliginous excrement and ingendered in those bladders. The tree does also yield his Turpentine by dropping like the former.

terebinthus.
The Turpentine tree.

terebinthus latifolia.
The broad leafed Turpentine tree.

The Place.

These trees grow, as Dioscorides says, in Iurie, Syria, Cyprus, Africke and in the Islands called Cyclades. Bellonius reports that there are found great store of them in Syria and Cilicia, and are brought from thence to Damascus to be sold. Clusius says that it grows of itself in Languedocke, and in very many places of Portingale and Spaine, but for the most part like a shrub and without bearing Turpentine.

Theophrastus writes that it grows about the hill Ida and in Macedonia, short, in manner of a shrub, and writhed; and in Damascus and Syria great, in manner of a small tree. He also sets down a certain male Turpentine tree and a female. The male, says he, is barren and the female fruitful. And of these he makes the one with a berry red at the first, of the bigness of a Lentill which cannot come to ripeness, and the other with the fruit greene at the first, afterwards somewhat of a yellowish red, and in the end blacke, waxing ripe in the spring, of the bigness of the Grecian Beane, and rosinny.

He also writes of a certain Indian Turpentine tree, that is to say, a tree like in boughes and leaves to the right Turpentine tree, but differing in fruit, which is like unto Almonds.

The Time.

The floures of the Turpentine tree come forth in the Spring together with the new buds. The berries are ripe in September and October, in the time of Grape gathering. The horns appear about the same time.

The Names.

The tree is called in Latine, *terebinthus;* in Italian, *terebintho;* in Spanish, *cornicabra;* in French, *terebinte;* in English, Turpentine tree. The Arabians call it *botin,* and with an article, *albotin.*

The Rosin is named in Latine, *terebinthina;* in high Dutch, *termintijn;* in English, Turpentine and right Turpentine; in the Arabian language, *albotin,* who name the fruit *granum viride,* or greene berries.

The Temperature and Vertues.

The barke, leaves and fruit of the Turpentine tree do somewhat bind, they are hot in the second degree and being green they dry moderately, but when they are dry they dry in the second degree, and the fruit approaches more near to those that be dry in the third degree, and also hotter. This is fit to be eaten, as Dioscorides says, but it hurts the stomacke.

It provokes urine, helps those that have bad spleens and is drunk in wine against the bitings of the poysonsome spiders called Phalangia.

The Rosin of the Turpentine tree excels all other Rosins, according to Dioscorides; but Galen writes that the Rosin of the Masticke tree bears the preheminence, and then the Turpentine.

This Rosin has also an astringent or binding faculty and yet not so much as the Masticke but it has withal a certain bitterness joined, by reason whereof it digests more than that of the Masticke tree. Through the same quality there is likewise in it so great a cleansing, as also it heals scabs, in his 8. book of the faculties of simple medicines, but in his book of medicines according to the kinds, he makes that of the Turpentine tree to be much like the Rosins of the Larch tree, which he affirms to be moister than all the rest, and to be without both sharpness and biting.

The fruit of Turpentine provokes urine and stirs up fleshly lust.

The Rosin of the right Turpentine, looses the belly, opens the stopping of the liver and spleen, provokes urine and drives forth gravell, being taken the quantitie of two or three Beanes.

The like quantity, washed in water divers times until it be white, then must be put thereto the like quantity of the yolk of an egge, and laboured together adding thereto by little and little (continually stirring it) a small draft of possit drink made of white wine, and given to drink in the morning fasting, it helps most speedily the Gonorrhæa, or running of the reines, commonly at the first time, but the medicine never fails at the second time of the taking of it, which gives stooles from four to eight, according to the age and strength of the patient.

Of the Wall-nut tree.

The Description.

nux juglans.
The Walnut tree.

This is a great tree with a thick and tall body. The bark is somewhat green and tending to the colour of ashes, and often times full of clefts. The boughs spread themselves far abroad. The leaves consist of five or six fastened to one rib, like those of the Ash tree, and with one standing on the top, which be broader and longer than the particular leaves of the Ash, smooth also and of a strong smell. The catkins or aglets come forth before the nuts. These nuts do grow hard to the stalk of the leaves by couples, or by three & three; which at the first when they be yet but tender have a sweet smell, and be covered with a green husk. Under that is a woody shell in which the kernel is contained, being covered with a thin skin, parted almost into four parts with a woody skin as it were; the inner pulp whereof is white, sweet, and pleasant to the taste; and that when it is new gathered, for after it is dry it becomes oily and rank.

The Place.

The Walnut tree grows in fields near common high ways, in a fat and fruitful ground, and in orchards. It prospers on high fruitful banks; it loves not to grow in watery places.

The Time.

The leaves together with the catkins come forth in the Spring. The nuts are gathered in August.

The Names.

The tree is called in Latin, *nux,* which name does signifiy both the tree and the fruit. In High Dutch, *nutzbaum;* in Low Dutch, *noote boome* and *nootelaer;* in French, *neisier;* in Spanish, *nogueyra;* in English, Walnut tree, and of some, Walsh Nut tree. The nut is called of some, *nux regia* or the Kingly Nut. It is likewise named *nux inglans.* In High Dutch, *baumnutz;* in Low Dutch, *walsch noten;* in Italian, *noci;* in French, *noix;* in English, Walnut and of some, Walsh nut.

The Temperature and Vertues.

The fresh kernels of the Nuts newly gathered are pleasant to the taste. They are a little cold and have no small moisture, which is not perfectly concocted. They be hard of digestion and nourish little. They slowly descend.

Dry Nuts taken fasting with a Fig and a little Rue withstand poison, prevent and preserve the body from the infection of the plague, and being plentifully eaten they drive worms forth of the belly.

The green and tender Nuts boiled in sugar and eaten as Suckad, are a most pleasant and delectable meat, comfort the stomach, and expel poison.

The oile of Walnuts made in such manner as oile of Almonds, makes smooth the hands and face, and take away scales or scurf, black and blew marks that come of stripes or bruises.

Milk made of the kernels, as Almond milk is made, cools and pleases the appetite of the languishing sick body. With onions, salt and honey, they are good against the biting of a mad dog or man, if they be laid upon the wound. Being both eaten and also applied, they heal in short time, as Disocorides says, gangrens, carbuncles, and the pulling away of the hair. This also is effectually done by the oile that is pressed out of them, which is of thin parts, digesting and heating.

Some of the later Physicians use these for baths and lotions for the body, in which they have a force to digest and also to procure sweat.

Of the Wayfaring tree.

The Description.

The Wayfaring man's tree grows up to the height of a hedge tree, of a mean bigness. The trunk or body thereof is covered with a russet bark.

The branches are long, tough, and easy to be bowed, hard to be broken, as are those of the Willow, covered with a soft whitish bark, whereon are broad leaves thick and rough, slightly indented about the edges, of a white colour, and somewhat hairy whilst they be fresh and green; but when they begin to wither and fall away, they are reddish, and set together by couples one opposite to another.

The flowers are white, and grow in clusters. After which come clusters of fruit of the bigness of a Pease, somewhat flat on both sides, at the first green, after red, and black when they be ripe. The root disperses it selfe far abroad under the upper crust of the earth.

The Place.

This tree grows in most hedges in rough and stony places, upon hills and low woods, especially in the chalky grounds of Kent, about Cobham, South Fleet, and Gravesend, and in all the tract to Canterbury.

The Time.

The flowers appear in Summer. The berries are ripe in the end of Autumne, and new leaves come forth in the Spring.

The Names.

This hedge tree is called *viurna* of Ruellius. In French, *viorne* and *viorna;* in Italian, *lantana.* It is reputed for the tree *viburnum,* of which Virgil makes mention in the first Eclog, where he commends the city of Rome for the loftiness and stateliness thereof, above other cities, saying that as the tall Cypres trees do show themselves above the low and shrubby Viorn, so does Rome above other cities lift up her head very high, in these verses:

Verum hæc tantùm alias inter cuput exulit urbes,
Quantum lenta solent inter virbuna cupressi.

But this all other cities so excels,
As Cypresse which 'mongst bending viornes dwells.

[Thomas Johnson writes: I judge *virbunum* not to be a name to any particular plant, but a general name to all low and bending shrubs, amongst which this here described may take place as one. I enquired of a country man in Essex, if he knew any name of this; he answered it was called the Cotton tree, by reason of the softness of the leaves.]

lantana, siue viburnam.
The Wayfaring tree.

The Temperature and Vertues.

The leaves and berries of Lantana are cold and dry, and of a binding quality.

The decoction of the leaves of Lantana is very good to be gargled in the mouth against all swellings and inflammations thereof, against the scurvie and other diseases of the gums, and fastens loose teeth.

The same boiled in lee does make the hairs black, if they be bathed or washed therewith, and suffered to dry of it self.

The berries are of the like facultie, the powder whereof when they be dried stay the laske, all issues of blood, and also the whites.

It is reported, that the bark of the root of the tree buried a certain time in the earth and afterward boiled and stamped according to art, makes good bird-lime for Fowlers to catch birds with.

Of the Wilding, or Crab tree.

malus sylvestris.
The Wilding or Crab tree.

The Kindes.

Like as there be diverse manured Apples, so are there sundry wilde Apples or Crabs, whereof to write apart were to small purpose, and therefore one description shall suffice for the rest.

The Generall Description.

There be diverse wilde Apple trees not husbanded, that is to say, not grafted; the fruit whereof is harsh and binding. For by grafting both Apples and Peares become more mild and pleasant. The Crab or Wilding tree grows often times to a reasonable greatness, equal with the Apple tree. The wood is hard, firm, and solid; the bark is rough, the branches or boughs many, the flowers and fruit like those of the Apple tree, some red, others white, some greater, others lesser. The difference is known to all, therefore it shall suffice what has been said for their several distinctions. We have in London gardens a dwarfe kind of sweet Apple, called *chamæmalus,* the dwarfe Apple tree or Paradise Apple, which bears Apples very timely without grafting.

The Place and Time.

The Crab tree grows in woods and hedgerows almost every where. The time answers those of the garden.

The Names.

Their titles do set forth their names in Latine and English: *malus sylvestris rubens,* the great wilding or red Crab tree; *malus sylvestris alba,* the white wilding or Crab tree; *malus sylvestris minor,* the smaller Crab tree; *malus duracina sylvestris,* the choking leane Crab tree.

The Temperature and Vertues.

Of the temperature of wild Apples hath been sufficiently spoken in the former chapter on the Apple tree.

The juice of wild Apples or Crabs takes away the heat of burnings, scaldings, and all inflammations, and being laid on in short time after it is scalded, it keeps it from blistering. The juice of Crabs (or verjuice) is astringent or binding, and has withal an abstersive or cleansing quality, being mixed with hard yeast of Ale or Beer, and applied in manner of a cold ointment, that is, spread upon a cloth first wet in the verjuice and wrung out, and then laid to, takes away the heat of Saint Anthonies fire, all inflammations whatsoever, heals scab'd legs, burnings and scalding wheresoever it be.

Of the Willow tree.

salix.

The common Willow.

The Description.

The common Willow is a high tree, with a body of a mean thickness, and rises up as high as other trees do if it be not topped in the beginning, soon after it is planted. The bark thereof is smooth, tough, and flexible. The wood is white, tough, and hard to be broken. The leaves are long, lesser and narrower than those of the Peach tree, somewhat green on the upper side, and slippery; and on the nether side, softer and whiter. The boughs be covered either with a purple or else with a white bark. The catkins which grow on the tops of the branches come first of all forth, being long and mossy, and quickly turn into white and soft down that is carried away with the wind.

The Rose Willow grows up likewise to the height and bigness of a shrubby tree, the body whereof is covered with a scabbed rough bark. The branches are many, whereupon do grow very many twigs of a reddish colour, garnished with small long leaves, somewhat whitish, amongst which come forth little flowers, or rather a multiplication of leaves, joined together in form of a Rose, of a greenish white colour; which do not only make a gallant show, but also yield a most cooling air in the heat of Summer, being set up in houses, for the decking of the same.

The dwarfe Willow has very small and slender branches, seldom times above a foot, but never a cubit high, covered with a duskish bark, with very little and narrow leaves, of a green colour above and on the upper side, but underneath of a hoary or overworn greenish colour, in bigness and fashion of the leaves of garden Flax. Among which come forth little duskish flowers, which do turn into down that is carried away with the wind. The root is small and thready, of a the bigness of a finger, and of a blackish colour.

The Place and Time.

These Willows grow in diverse places of England. The Rose Willow grows plentifully in Cambridgeshire, by the rivers and ditches there in Cambridge town they grow abundantly about the places called Paradise and Hell-mouth, in the way from Cambridge to Grandchester. I found the dwarf Willows growing near to a bog or marish ground, at the further end of Hampsted Heath upon the declining of the hill, in the ditch that encloses a small cottage there, not half a furlong from the said house or cottage.

The Willows do flower at the beginning of Spring.

The Names.

The Willow tree is called in Latin, *salix;* in Low Dutch, *wilgen;* in Italian, *salice* or *salcio;* in French, *saux;* in Spanish, *salgueiro;* in English, Sallow, Withie, and Willow. The greater is called in Latin, *salix perticalis,* especially that which being often lopped sends out from one head many boughs.

The Temperature and Vertues.

The leaves, flowers, seed and bark of Willows are cold and dry in the second degree, and astringent. The leaves and bark of Withy or Willows do stay the spitting of blood, and all other fluxes of blood whatsoever in man or woman, if the said leaves and bark be boiled in wine and drunk.

The green boughs with the leaves may very well be brought into chambers and set about the beds of those that be sick of fevers, for they do mightily cool the heat of the air, which thing is a wonderful refreshing to the sick patients.

The bark has like vertues. Dioscorides writes that this being burnt to ashes, and steeped in vinegar, takes away corns and other risings in the feet and toes. Diverse, says Galen, do slit the bark while the Withy is in flowering, and gather a certain juice, with which they use to take away things that hinder the sight, and this is when they are constrained to use a cleansing medicine of thin and subtle parts.

Of the Yew tree.

The Description.

The Yew that bears Acorns and berries is a great high tree remaining always green, and has usually a huge trunk or body as big as the Oke, covered over with a scabbed or scaly bark. The timber hereof is somewhat red, near as hard as Box, universally covered next the bark with a thick white sap like that of the Oke. It has many big limbs, divided into many small spreading branches. The leaves be about an inch long, narrow like the leaves of Rosemary, but smooth and of a darker green colour, growing all along the little twigs or branches close together.

The Place.

This tree is very common in England. In Hampshire, there is a good plenty of them growing wild on the chalky hills, and in church yards where they have been planted.

taxus.

The Yew tree.

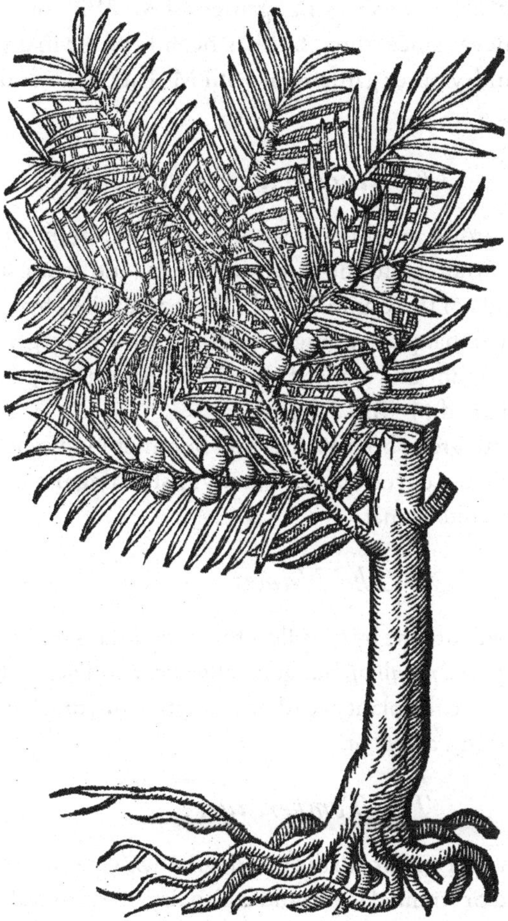

The Names.

It is named in Latin, *taxus;* in High Dutch, *eybenbaum;* in Italian, *tasso;* in Spanish, *toxo* and *taxo*; in English, Ewe or Yew tree. In the unlearned shops of Germany, if any of them remain, it is called *tamariscus;* where in times past they were wont, not without great error, to mix the bark hereof in compound medicines, in stead of the Tamariske bark.

The Nature.

All that heretofore have dealt in the faculty of Herbarisme, say that the Yew tree is very venomous to be taken inwardly, and that if any do sleep under the shadow thereof, it causes sickness and oftentimes death. Moreover, they say that the fruit thereof, being eaten, is not only dangerous and deadly unto man, but if birds do eat thereof, it causes them to cast their feathers and many times to die. All of which I dare boldly affirm is altogether untrue; for when I was young and went to school, diverse of my school fellows and likewise my self, did eat our fill of the berries of this tree and have not only slept under the shadow thereof, but among the branches also, without any hurt at all, and that not one time, but many times.

Of Yucca or Iucca.

The Description.

The plant of whose root the Indian bread called *cazava* is made, is a low herb consisting only of leaves and roots. It has neither stalk, flowers, nor fruit, that I can understand of others, or by experience of the plant it self, which has grown in my garden four years together, and yet does grow and prosper exceedingly, nevertheless without stalk, fruit, or flowers as aforesaid.

It has a very great root, thick and tuberous, and very knobby, full of juice somewhat sweet in taste, but of a pernicious quality. From which root rises up immediately forth of the ground very many leaves joined unto the head of the root in a round circle; the which are long, of the length of a cubit, hollowed like a gutter or trough, very smooth, and of a green colour like that of Woade. The edges of the leaves are sharp like the edge of a knife, and of a brown colour. The point of the leaf is a prick as sharp as a needle, which hurts those that unadvisedly pass by it.

yucca, siue iucca peruana.
The root whereof the bread
casava or cazava is made..

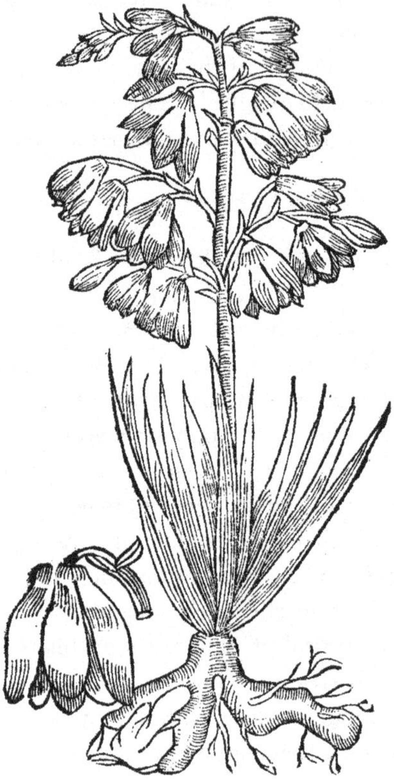

The leaf is like unto a little wherry, or such like boat. They are also very tough, hard to break, and not easy to cut, except the knife be very sharp.

I saw it once flower in the garden of Mr Wilmot at Bow, but never since, though it has been kept for many years in sundry other gardens, as with Mr Parkinson and Mr Tuggie.

The Place and Time.

This plant grows in all the tract of the Indies, from the Magellane Straits unto the Cape of Florida, and in most of the Islands of the Canibals and others adjoining, from whence I had that plant brought me that grows in my garden, by a servant of a learned and skilful Apothecary of Excester, named Mr Tho. Edwards.

It keepeth green both Winter and Summer in my garden without any coverture at all, notwithstanding the injurie of our cold clymat.

The Names.

It is reported unto me by travellers that the Indians do call it in some parts, *manihot,* but generally *yucca* and *iucca.* It is thought to be the plant called of Theophrastus, *arachidna,* and of Pliny, *aracidna.*

The Temperature.

This plant is hot and dry in the first degree, which is meant by the feces or dross, when the poisonous juice is pressed or strained forth, and is also dry in the middle of the second degree.

Of the Goose tree, Barnacle tree, or the tree bearing Geese.

The Description.

Having travelled from the grasses growing in the bottom of the fenny waters, the Woods, and mountains, even unto Libanus it self; and also the sea and bowels of the same, we are arrived at the end of our Historie; thinking it not impertinent to the conclusion of the same, to end with one of the marvels of this land (we may say of the world). The history whereof to set forth according to the worthiness and rarity thereof, would not only require a large and peculiar volume, but also a deep research into *britannicae conchae anatiferae.*

The breed of barnacles.

the bowels of nature, than my intended purpose will suffer me to wade into, my sufficiency also considered; leaving the history thereof rough-hewn, unto some excellent men, learned in the secrets of Nature, to be both fined and refined. In the mean space, take it as it falls out: the naked and bare truth, though unpolished.

There are found in the North parts of Scotland and the Islands adjacent, called Orchades, certain trees whereon do grow certain shells of a white colour tending to russet, wherein are contained little living creatures, which shells in time of maturity do open, and out of them grow those little living things, which falling into the water do become fowls, which we call Barnacles; in the North of England, brant Geese; and in Lancashire, tree Geese; but the other that do fall upon the land perish and come to nothing. Thus much by the writings of others, and also from the mouths of people of those parts, which may very well accord with truth.

But what our eyes have seen and hands have touched, we shall declare. There is a small Island in Lancashire called the Pile of Foulders, wherein are found the broken pieces of old and bruised ships, some whereof have been cast thither by shipwracke, and also the trunks and bodies with the branches of old and rotten trees, cast up there likewise; whereon is found a certain spume or froth that in time breeds unto certain shells, in shape like those of the Muskle, but sharper pointed, and of a whitish colour; wherein is contained a thing in form like a lace of silk finely woven as it were together, of a whitish colour, one end whereof is fastened unto the inside of the shell, even as the fish of Oisters and Muskles are. The other end is made fast unto the belly of a rude mass or lump, which in time comes to the shape and form of a Bird. When it is perfectly formed the shell gapes open, and the first thing that appears is the foresaid lace or string; next come the legs of the bird hanging out, and as it grows greater it opens the shell by degrees, till at length it is all come forth, and hangs only by the bill. In short space after it comes to full maturity and falls into the sea, where it gathers feathers, and grows to a fowl bigger than a Mallard and lesser than a Goose, having black legs and bill or beak, and feathers black and white, spotted in such manner as is our Mag-Pie, called in some places a Pie-Annet, which the people of Lancashire call by no other name than a tree Goose; which place aforesaid, and all those parts adjoining do so much abound therewith, that one of the best is bought for three pence. For the truth hereof, if any doubt, may it please them to repair unto me, and I shall satisfy them by the testimony of good witnesses.

Moreover, it should seem that there is another sort hereof; the History of which is true, and of mine own knowledge: for travelling upon the shore of our English coast between Dover and Rumney, I found the trunk of an old rotten tree, which (with some help that I procured by Fishermen's wives that were there attending their husbands' return from the sea) we drew out of the water upon dry land. Upon this rotten tree I found growing many thousands of long crimson bladders, in shape like unto puddings newly filled, before they be sodden, which were very clear and shining; at the nether end whereof did grow a shell fish, fashioned somewhat like a small Muskle, but much whiter, resembling a shell fish that grows upon the rocks about Garnsey and Garsey, called a Lympit. Many of these shells I brought with me to London, which after I had opened I found in them living things without form or shape; in others which were nearer come to ripeness I found living things that were very naked, in shape like a Bird. In others, the Birds covered with soft down, the shell half open, and the Bird ready to fall out, which no doubt were the Fowls called Barnacles. I dare not absolutely avouch every circumstance of the first part of this history, concerning the tree that bears those buds aforesaid, but will leave it to a further consideration; howbeit, that which I have seen with mine eyes, and handled with mine hands, I dare confidently avouch, and boldly put down for verity.

Now if any will object that this tree which I saw might be one of those before mentioned, which either by the waves of the sea or some violent wind had been overturned as many other trees are; or that any trees falling into those seas about the Orchades, will of themselves bear the like Fowles, by reason of those seas and waters, these being so probable conjectures and likely to be true, I may not without prejudice gainsay or endeavour to confute.

The Place and Time.

The borders and rotting planks whereon are found these shells wherein is bred the Barnacle, are taken up in a small island adjoining to Lancashire, half a mile from the maine land, called the Pile of Foulders. They spawn, as it were, in March and April. The Geese are formed in May and June, and come to fullness of feathers in the month after.

And thus having through God's assistance discoursed somewhat at large of grasses, herbs, shrubs, trees, and mosses, and certain excrescences of the earth, with other things more, incident to the history thereof, we conclude and end our present Volume, with this wonder of England. For the which God's name be ever honoured and praised.

Finis.

Made in the USA
Monee, IL
08 November 2026

67967207R00098